THE ANTHROPOLOGY OF DISEASE

BIOSOCIAL SOCIETY SERIES

Series editor: G. A. Harrison

THE
ANTHROPOLOGY
OF DISEASE

Edited by

C. G. N. MASCIE-TAYLOR

*Department of Biological Anthropology
Cambridge*

Oxford New York Tokyo
OXFORD UNIVERSITY PRESS
1993

Oxford University Press, Walton Street, Oxford OX2 6DP
Oxford   New York   Toronto
Delhi   Bombay   Calcutta   Madras   Karachi
Kuala Lumpur   Singapore   Hong Kong   Tokyo
Nairobi   Dar es Salaam   Cape Town
Melbourne   Auckland   Madrid
and associated companies in
Berlin   Ibadan

Oxford is a trade mark of Oxford University Press

Published in the United States
by Oxford University Press Inc., New York

© The contributors listed on p.xi, 1993

A catalogue record for this book is available from the British Library

Library of Congress Cataloging in Publication Data
The Anthropology of disease/edited by C. G. N. Mascie-Taylor.
(Biosocial Society series: 5)
1. Medical anthropology.   2. Diseases—Social aspects.
3. Environmentally induced diseases.   I. Mascie-Taylor, C. G. N.
II. Series.
[DNLM: 1. Anthropology, Cultural.   2. Disease—ethnology.
GN 320 A628 1993]
GN296.A628   1993   306.4'61—dc20   93–3221

ISBN 0–19–852287–8 (hbk.)

Typeset by
Advance Typesetting Ltd, Oxford
Printed in Great Britain by
Biddles Ltd, Guildford & King's Lynn

PREFACE

This book is the fifth in a series that the Biosocial Society of Great Britain has produced in conjunction with Oxford University Press. The aim of the society is to promote studies of biosocial matters and to examine topics and issues which have both biological and social importance.

The present volume brought together three contributors who all research on different aspects of disease. Nick Mascie-Taylor studies the biology and epidemiology of schistosomiasis, ascariasis, and other worms, and his chapter concentrates on the biological anthropology of disease. Gilbert Lewis has studied the social perspective of disease in Papua New Guinea and Africa and he discusses some of the social anthropological elements. Finally, Stephen Kunitz who has wide knowledge and interest in indigenous peoples, including American Indians, provides a demographic case study of the impact of European contact on the mortality patterns in Polynesia.

All three contributors were able to address a meeting of the Biosocial Society held in May 1990 at the Pauling Centre, University of Oxford. The society is most grateful to Professor G. A. Harrison for the local arrangements including making the Centre available for the meeting.

Finally, grateful acknowledgement is made to the Parkes Foundation for a grant which made this monograph possible.

Cambridge   C.G.N.M-T
1993

CONTENTS

CONTRIBUTORS

C. G. N. Mascie-Taylor
Head of the Department of Biological Anthropology, University of Cambridge, and Fellow of Churchill College, Cambridge, UK

G. Lewis
Lecturer in Social Anthropology, University of Cambridge, and Fellow of St John's College, Cambridge, UK

Stephen J. Kunitz
Professor in the Department of Community and Preventive Medicine, University of Rochester Medical Center, Rochester, USA

1

THE BIOLOGICAL ANTHROPOLOGY OF DISEASE

C. G. N. *Mascie-Taylor*

INTRODUCTION

The study of disease in past and present-day human populations provides insight into the relationship between people and their environments. Changing disease patterns reflect the net effect of differences in demography, climate, nutrition, vegetation, dietary habits, religion, culture, behaviour, migration, occupation, economic exploitation, social structure, and genetic make up of populations. This chapter focuses on how biocultural factors influence disease patterns and on the biological impact of disease on past and present-day human populations.

SECTION 1: BASIC PRINCIPLES

Definition of disease

Disease can be defined as an 'impairment of health and well-being'. Indeed, good health would be freedom from disease. Conceptually good health is easy to define; properly integrated functioning of all bodily components to give optimal function of the total individual—mental, emotional, and physical. In practice, 'optimal' is not definable and proof of good health status comprises a relatively subjective evaluation of the overall functional status of the individual. Individuals are considered to be healthy if they fall within the wide limits of normal functioning set by observation of normal people.

However, societies vary in what and how they define 'normal'. Thus the cultural anthropologist looks to see how a society perceives and defines disease (see Chapter 2). This approach, sometimes called the ethnomedical, has been more often used in non-Western societies but is receiving increasing use in Western countries. For instance, the ethnomedical approach has been used to study AIDS (acquired immune deficiency syndrome) and mental behaviour (Armelagos et al. 1990).

Biological anthropologists, on the other hand, study human adaptability, the basic biological flexibility of human populations. The study of human adaptability, which embraces all of the mechanisms by which human populations adjust to their environment forms the inter-disciplinary field of human ecology. Human ecology embraces the disciplines of environmental physiology, growth and nutrition, epidemiology, human genetics, and demography.

Biological anthropologists base their study of disease on the 'ecological model' which is derived from epidemiology (Armelagos *et al.* 1990). This model views disease as involving the interaction between the agent (the animate or inanimate proximal cause), the human host, and the environment. The biological anthropology dimension extends this model to analyse the biocultural response and impact of the disease.

What causes disease?

It is an illusion that the primary cause is the only important cause of a disease as can be illustrated for tuberculosis. Humans contract this disease because they become the involuntary host of the agent, the tubercle bacillus. In Denmark during World War II, 300 students were exposed in a poorly ventilated room to a teacher who developed open tuberculosis. Of the student group, 94 had not been vaccinated nor had any natural infection as determined by a skin reaction test. Yet 24 escaped infection and remained tuberculin-negative; 29 experienced subclinical infection; of the remaining 41 students who showed evidence of primary tuberculosis only 14 developed progressive pulmonary disease.

Simply pointing to the tubercle bacillus as the primary cause does not fully answer the question 'Why is it that only some humans get tuberculosis?'—otherwise all 94 rather than just 41 students would have shown primary tuberculosis. There must be important additional factors which need to be considered.

Steps in the development of disease

The easiest way of appreciating the process by which a disease ensues is to question what steps lead to the disease. Did sputum or nasal droplets carrying the tubercle bacillus reach all the students? It depends on the number of sources of the bacillus (the teacher and previously affected students), their proximity to the individual, and the actions taken by individuals to avoid exposure.

Even if the bacillus reaches an individual does it infect him or her? A large number of factors in the environment have to be taken into account: the abundance of bacilli in the sputum source which may or may not be adequate to cause an infection; the influence of sunlight and other environmental variables which kill tubercle bacilli; and the route of entry into the individual may or may not provide access to susceptible tissue (in the case of the tubercle bacillus, entry must be respiratory or oral, not dermal). Finally, the individual's defence system might prevent multiplication of the bacilli.

Even if the infection becomes established in the individual he or she might not show clinical signs of the disease. Such factors as dosage, pathogenicity, resistance to disease, including genetic influences, nutritional status, and specific immunity induced by prior natural infection or BCG vaccine all have an influence.

It can therefore be seen that causation is rarely single or simple. A whole complex of causes (defined as the 'causal web' by Dunn and Janes 1986) is constantly at work in the individual, in the group, and in the environment. The individual causes include various stress factors which affect resistance, a latent infecting agent perhaps acquired in childhood, genetic determinants of susceptibility or resistance, and habits and patterns of life. The group comprises the source of infection (both cases and carriers). Other causes include the overall density of the population as it affects the frequency and extent of human contact and the degree of co-operation to combat adverse environmental influences. In the environment it is well known that the tubercle bacillus can be perpetuated in cattle and there are many other factors (see later) which affect human exposure and susceptibility.

In summary, even for a single disease, such as tuberculosis, the determinants include: the tubercle bacillus and its strain-determined pathogenicity; the genetic character, the age, sex, and nutritional status of the host; the operation of various stress phenomena; the extent of human contact in homes and places of work; the adequacy of ventilation and sunlight; and the availability of medical care and proper nutrition.

Classification of disease agents

So far we have dealt with one disease and its agent, the tubercle bacillus. In other diseases the agent can be any organism, substance, or a force, the excessive presence or relative lack of which is the immediate or proximal cause. For many diseases, including coronary heart disease and peptic ulcer, the agent so defined is still unrecognized. Several classifications of agents have been proposed. One is given in Table 1.1.

The host–agent interaction

The long-term survival of a parasite depends on its viability (the ability to survive in the free state), on its growth requirements, on its host range (the

TABLE 1.1. A classification of agents of diseases

1. *Nutritional elements*. Lack or over-abundance of a specific food substance or class of substances, such as fats, carbohydrates, proteins, vitamins, minerals, can result in disease.
2. *Chemical agents*. These comprise poisons, allergens, and irritants, which arise outside the host and are capable of producing disease or normal breakdown products of tissue metabolism which accumulate because kidney dysfunction interferes with their elimination.
3. *Physiological factors*. Events and changes in the normal life span can give rise to specific disease conditions in certain individuals. Morning sickness in pregnancy is a good example.
4. *Genetic factors*. Nowadays a large number of genetic diseases are known ranging from point mutations to chromosomal anomalies.
5. *Psychic factors*. The more common situations in which psychic factors are believed to play a part include headaches, nausea, vomiting, peptic ulcers, and hypertension.
6. *Physical factors*. These factors include fire, sunlight, and gravity changes leading to a range of diseases including the 'bends'.
7. *Invading living organisms*. These comprise one of the largest groups of disease agents. Invading organisms belonging to the animal kingdom are called 'higher parasites' (or parasites for short) and are either metazoan (multicellular animals), such as arthropod mites of scabies and helminths (intestinal worms), or protozoan (single celled animals), such as amoeba and the malarial parasites. Other organisms that serve as agents of disease are fungi, bacteria, rickettsiae, and viruses.

spectrum of animals and arthropods an agent can successfully parasitize or infect), its vulnerability to chemotherapeutic or antibiotic substances, the extent to which it can change its antigenic character, and whether it can adapt to a new reservoir.

Three concepts are used to describe the response of individuals to microbial agents of disease: infectivity, pathogenicity, and virulence.

Infectivity is the property of being able to lodge and multiply in a host, i.e. to infect a host. The basic measure of infectivity is the minimum number of infective particles required to establish an infection. This number, which may vary substantially from one host to another, depends on such factors as entry route, host age, and other host characteristics.

Pathogenicity refers to the ability of microbial agents to induce disease. This ability depends upon the rapidity and degree to which the agent multiplies, the extent of tissue damage caused by the multiplication and whether or not the agent produces a specific toxin (e.g. diphtheria and tetanus bacilli). The measure of pathogenicity is simply the proportion of infections which result in disease.

Virulence refers to the severity of the disease. Criteria of severity may be permanent and serious sequelae or death. The measure of virulence is the number of severe cases over the total number of cases.

Diseases vary in their infectivity, pathogenicity, and virulence. Smallpox is an example of a disease which has high infectivity, pathogenicity, and virulence, whereas chickenpox has high infectivity and pathogenicity but low virulence. On the other hand leprosy has very low infectivity and pathogenicity but high virulence.

Modes of transmission

For many organisms including measles, mumps, and the poliomyelitis virus, humans are the only host, that is there is only one reservoir. For a few, such as yellow fever where the common host is a monkey, humans are an important alternate host. Infections of invertebrate hosts are important to the reservoir mechanism as the infection of the arthropod may be necessary for the completion of an essential stage of the development cycle of the parasite.

Three steps: escape from the source host; conveyance to; and effective entry into the recipient host, make up the transmission cycle. Escape means the emergence of an agent from an infected source host. The more usual avenues of escape are discharges from infected lesions; saliva; mucous secretions of the respiratory tract and secretions of the reproductive system; and urine and faeces. Another important avenue is the bloodstream by way of cuts, hypodermic needles, transfusions of bites of insects or other animals.

Having escaped from an infected host, the agent has to move to a susceptible host. There are a number of ways in which conveyance can take place. Direct contact involves no vector and it occurs in kissing, shaking hands, sexual intercourse, and short-range airborne spread. Contagious diseases are those spread by direct contact and include smallpox, measles, and chickenpox.

Indirect modes of conveyance involve a vector. The vector can be either animate or inanimate. Examples of inanimate vectors are infected water droplets, fomites (intimate personal articles such as handkerchiefs), food, milk, and water. Animate vectors are either mechanical or biological. A good example of a mechanical vector is the fly which carries animal excrement and deposits it on some exposed foodstuff. To be classed as a biological vector an animal must experience infection with multiplication of the microbial agent before transmission can occur. In louse-borne typhus, for example, the lice eventually die of the infection. Malaria, yellow fever, and schistosomiasis are other diseases which involve biological vectors.

Finally, the agent has to gain entry to the host. There are a number of entry points including through broken skin, the respiratory tract, the alimentary canal, the bloodstream, or through injection through intact skin by blood-sucking arthropods.

Host defence systems

Contact with an agent may result in one of three possible events. First, because of inadequate dosage or unsuitable portal of entry or specific host immunity, the parasite fails to lodge and establish an infection. Secondly, the infection may be established but remain subclinical. Thirdly, the infection may cause disease.

The host has several lines of defence. Structurally, the intact skin and mucous membranes provide barriers which resist penetration. Functional aspects are also important. Coughing and sneezing tend to rid the respiratory passages of harmful substances. Pain, touch, smell, taste, sight, and hearing all activate evasive action if danger threatens.

The immunological response is triggered by the presence, in the host, of antigens which it recognizes as foreign or non-self. Usually this results in the production of protective antibodies. These are a response to the infection but may already be present from naturally acquired active immunity that results from previous infection. Artificially acquired active immunity is stimulated by administration of vaccines (or toxoids).

Newborn infants possess naturally acquired passive immunity to many agents. They acquire this short-lived immunity (lasting about 25 days) mainly by transplacental passage of maternal antibodies into the fetal circulation. Further immunity is acquired via maternal milk.

Artificially acquired passive immunity also exists and this is most frequently conferred through use of gamma globulin.

SECTION 2: FACTORS MODIFYING DISEASE PATTERNS

Socio-cultural evolution

The development of agriculture led to significant changes in human habits which heralded a dramatic increase in infectious diseases (Fenner 1980). Two changes are particularly important: first, people started living in settled aggregates which became larger and more densely populated over time; and secondly, there was increased proximity and contact between humans and other animals (Table 1.2).

Reconstructing the disease profiles of early hominids has been attempted (Table 1.3). It is generally believed that human viral diseases, such as

TABLE 1.2. Cultural characteristics in relation to the number of human generations and population aggregation

Years before 1985	Generations	Cultural state	Size of human communities
1 000 000	50 000	Hunter and food gatherer	Scattered nomadic bands of <100 persons
10 000	500	Development of agriculture	Relatively settled villages of <300 persons
5500	220	Development of irrigated agriculture	Few cities of 100 000; mostly villages of <300 persons
250	10	Introduction of steam power	Some cities of 500 000; many cities of 100 000; many villages of 1000 persons
130	6	Introduction of sanitary reforms	–
0	–	–	Some cities of 5 000 000; many cities of 500 000; fewer villages of 1000

measles, would have been absent from hunter-gatherers because of the size of community required to maintain the measles virus. Hunter-gatherers are generally thought to have existed in small nomadic bands with limited contact between groups. The measles virus can persist in one person for about two weeks, from the time of infection to the appearance of neutralizing antibodies (Black 1980). The virus must therefore move at least 26 times a year and can only survive in a population which produces that number of babies. It is most unlikely that nomadic bands would have provided babies at such regular intervals! The same principle of a minimum size of the population would be true for other viral diseases with no animal reservoir (Fenner 1980). The only viral diseases that would be expected in these societies would have been those with recurrent infectivity and latency, like chickenpox and herpes simplex.

Polgar (1964) suggests that hunter-gatherers would have two types of disease to contend with: (1) those adapted to pre-hominid ancestors, such as head and body lice, pinworms, yaws, and possibly malaria; and (2) zoonoses which are diseases which have non-human animals as their primary host. Examples of such diseases are plague, flea- and mite-borne typhus, and sleeping sickness. Livingstone (1958) discounts the likelihood of early

TABLE 1.3. Disease profiles, early hominids to the present

	Present	Absent
Hunter-gatherer	Arbovirus, chickenpox, rabies, tuberculosis, herpes simplex	Human viral diseases, some bacterial infections, e.g. cholera, typhoid
Agriculture 1. Primitive villages	All those found in Hunter-gatherers + Enteric bacteria + Respiratory infections	Measles, smallpox, rubella
2. Primitive cities	All diseases with human–human spread	Measles, smallpox, rubella
3. Advanced cities	Measles, rubella, venereal diseases	Due to controls, e.g. clean water, vaccination, chemotherapy

hominids having malaria as they lived in savannah and because of their small population sizes. Both Fenner (1980) and Cockburn (1967a, b) would also include bacteria, such as salmonella, streptococci, and staphylococci, because these bacteria can survive indefinitely on skin or mucous membranes and invade the body when the surface is damaged and exposed.

The development of agriculture was, unfortunately, not allied with an understanding of sanitation. Contact with human waste would have led to the development of endemic enteric bacterial infections and herding practices would increase the frequency of contact with zoonotic diseases. As noted by Armelagos et al. (1990) products of domesticated animals such as milk, hair, and skin, as well as dust, could transmit anthrax, brucellosis, and tuberculosis. However, it is likely that the generalized viral infections and most of the enteric and respiratory viruses failed to become established (Fenner 1980). Agricultural activities and increased population size may have increased the transmission of malaria. Zulueta (1956) showed that nomadic hunter-gatherers had lower malarial parasite and spleen rates than slash and burn agriculturists living in the same forest areas of Sarawak (East Malaysia). In addition the destruction of the forest created more favourable conditions (drier and sunnier) for the anopheline mosquitoes, the carriers of malaria, to breed.

Irrigation made large-scale agriculture possible and as the settlements grew in size, becoming towns and eventually cities, the population size became sufficient to maintain viral infections, such as measles, smallpox, and rubella. In addition, the high population density enabled the spread of respiratory and faecal–oral viruses. Sanitary conditions deteriorated and the lack of potable water increased the likelihood of the spread of cholera.

The physical environment

The contribution of the physical environmental factors to disease patterns is difficult to evaluate for a number of reasons. First, there is the multiplicity of factors. Secondly, factors typically operate concurrently and in an interactive way. In addition, the effects may be indirect and some environmental factors have the potential to act on the agent and the host and on the agent–host relationship. Sunlight is a good example of a factor which has a 'triple-barrelled' function. For the tubercle bacillus, ultraviolet light may be lethal whereas sunlight is a source of bodily warmth in humans. Sunlight enables humans to synthesize vitamin D, but it is also a disease agent causing sunburn and promoting skin cancer.

In hot dry climates sweating is the main way of losing heat. If sweat does not evaporate it stays on the skin, raises the pH and allows microorganisms to multiply. In places experiencing a rainy season, sweat evaporates slowly, *Streptococcus pyogenes* and *Staphylococcus aureus* may flourish. In one area of Ghana, nearly 20 per cent of the population had pyoderma (bacterial diseases in which pus is formed) at some time of the year. Farmers and labourers, who sweat a great deal, were particularly vulnerable. Fungi also flourish when the skin is wet and fungal infections get worse in the rainy season. One of the risks of prolonged residence in hot climates may be urinary stones. There is no simple relationship between water shortage and stone formation, and it has been suggested that rubidium in the diet may help stones to form. Urinary stones are found particularly in the Sahelian countries of Africa.

Considerable attention has been paid to the effects of climate and seasonality on vectors and agents of disease. Hot climates favour many of the vectors of common diseases. Two examples illustrate the importance of rainfall and temperature. The development of infective larvae of *Onchocerca volvulus* (causing river blindness) within the fly *Simulium damnosum* or *S. naevi* depends largely on external temperature, which is optimal at 24 °C. Larvae do not develop at temperatures below 18 °C. Breeding is impossible during the dry season. The multiplication of *Trypanosoma gambiense* (causing sleeping sickness) in tsetse flies can only take place at temperatures of 24 °C and above and temperatures between 25 °C and 30 °C are optimal. Extreme dryness or much wetness are lethal to the flies and a relative humidity of 40–60 per cent is required. Temperatures of 35 °C and above are also lethal and few flies can survive above 40 °C. As such temperatures are common in the savannah during the dry season, sleeping sickness will only be found in those places where the flies can survive, i.e. protected by shade from trees and near water.

The best example of a parasite which depends very largely on climate for its breeding is the malarial parasite. Temperatures lower than 20 °C indefinitely delay the development of *Plasmodium falciparum*. At 30 °C the development of this malarial parasite within the mosquito takes 9 days whereas at 20 °C the process takes 20 days. The virus causing yellow fever takes 4 days to mature in the mosquito vector at an ambient temperature of 37 °C but takes 36 days at 18 °C. Epidemics of yellow fever are therefore more likely to occur when the temperature is high enough for rapid multiplication and transmission.

Seasonal effects

Seasonal changes in transmission of infection occur. Three distinct patterns can be recognized: (1) human–vector or human–intermediate host contact is increased; (2) humans become more susceptible to the pathogen; or (3) the pathogen spreads more easily and so reaches more people. If the mosquito cannot breed, as in places with a long dry season, then continuous transmission of yellow fever cannot occur. Intermediate hosts like the crustacean *Cyclops* which harbours Guinea worm larvae are washed, in the torrential rainy season of the forest, away from the surface water, so infected water is not regularly drunk, and transmission of the disease decreases. However, in the savannah, the early rains fill surface pools from which people collect water. These pools are a common habitat of *Cyclops*.

The very hot season in the Sudan prevents reproduction of snails of the *Biomphalaria* species, which are intermediate hosts in the life cycle of schistosomiasis, thereby limiting its distribution. The reproductive activity of the snail is also limited by cold so that in Egypt during the winter, and in the cold season elsewhere, there are fewer snails to act as intermediate hosts and therefore the incidence of new infections decreases.

Airborne infections by bacteria and viruses also show seasonal patterns. Measles is epidemic during the dry season particularly during the colder months. Meningococcal meningitis also shows a seasonal pattern with large outbreaks every year towards the end of the dry season.

Contact with vectors or organisms shows seasonal patterns. It has already been noted that the tsetse fly cannot withstand high temperatures and so in the dry season breeds close to water. Humans also move towards streams or rivers at this time to get water. Thus, there is increased contact by humans with the flies carrying sleeping sickness when humans collect drinking water, water their cattle, wash, or cultivate a narrow strip of irrigated land. Onchocerciasis also shows a peak in the dry season for similar reasons.

The scarcity of food towards the end of the dry season may cause under-nutrition. Folic acid deficiency varies with the supply of fresh vegetables and root crops. In many parts of the world poorer people have to go through a hungry period before the harvest season when they are likely to be particularly susceptible to infection.

Movement patterns

Migration has been and continues to be one of the major demographic processes affecting world populations. People migrate for a multitude of reasons including to find work, escape oppression, and for social reasons. Kaplan (1988) has provided a comprehensive review of the relationship between migration and disease, covering both infectious and non-infectious diseases.

Historically, three new patterns of travel exposed large and dense civilized populations to new infections for the first time (McNeill 1976). The earliest occurred in the first Christian centuries when caravans and ships linked China and India with the Mediterranean coastline. McNeill suggests that the new and lethal infections were partly responsible for the collapse of the Roman and Chinese (Han) empires. The second new pattern was associated with the establishment of the Mongol empire in the thirteenth century. Genghis and his heirs succeeded in uniting most of northern Eurasia. This created a ramshackle network, the probable side-effect of which was to effect the transfer of *Yersinia pestis* (the bacillus for the bubonic plague) via the rats in the saddlebags of the caravanserais from a region of endemicity among burrowing rodents in the Burma–Yunnan border to the burrowing rodents of the Eurasian steppe. As McNeill (1976) has put it: 'the microparasites accompanying the macroparasites'.

Arrival of infected rats in the Eastern Mediterranean coincided with the opening of the waterways for trade (the third new pattern), the return of the Crusaders, and recent crop failures so that people had moved into already crowded cities. The greatest epidemic of any disease of any time—the Black Death—commenced in the Middle East in 1346 and as Fig. 1.1 shows, it swept across Europe in four years. It has been estimated that about 40 million people died, approximately one-quarter of Europe's population.

The results of the introduction into a community of a disease where it has not been endemic have usually been catastrophic. For example, European colonization led to severe epidemics of measles being introduced to the people of the Amazon in 1749, to Estonia in 1829, to the Indians of Hudson Bay territory in 1846, to the Hottentot of the Cape in 1852, to Tasmania in 1854. The population of Tierra del Fuego was almost wiped out by measles in 1883.

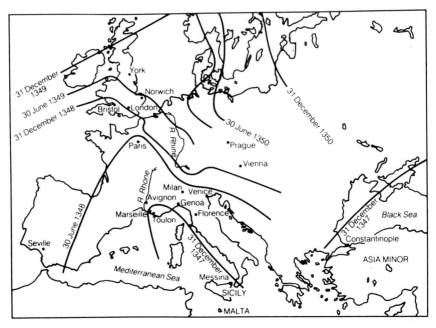

Fig. 1.1   Areas affected by the Black Death in Europe, 1348–50. (Open University 1985.)

The first introduction of measles into Fiji in 1875 resulted in 30 per cent of the population dying within three months. Partly as a result of this measles epidemic, Indian labour was introduced to work on the plantations. The whole culture of the island changed and the recent civil unrest has largely been ascribed to the simmering disagreements between the Fijian and Indian communities.

Smallpox had a major impact decimating populations and contributed to the conquest of the Incas, Aztecs, and North American Indians. As a result of trade with Africa, ships carrying slaves and water casks containing *Aedes aegypti* and yellow fever virus, there were continuous outbreaks of yellow fever for more than 250 years (Stanley 1980). For example, yellow fever epidemics occurred in 56 different years and on 37 occasions in North America in the nineteenth century and it wiped out the French Army of 25 000 sent by Napoleon Bonaparte to Santo Domingo.

In more recent times major infections and epidemics have often followed trade routes. For example, the spread of louse-borne relapsing fever across western Sudan after the two world wars was following a trade route; and plague in Ghana in 1908 and 1924 was spread by people moving from village to village along the coast. Modern air transport means that short-lived organisms can now be transmitted.

Customs and habits

Traditional customs may lead to disease and modern habits may affect health. How food is prepared and stored is particularly important. Badly stored food may result in fungal growth and the production of toxins. Some parts of food are harmful. The skin of cassava contains cyanide and, when it is boiled and pounded in with the rest of the tuber, cyanide is released which gives rise to ataxic neuropathy. Allergic conjunctivitis and laryngitis are common in the peoples of northern Sudan and are thought to be due to their habit of eating raw liver, lungs, and intestines of goats which are heavily infected with parasites.

Kuru is a chronic progressive disease of the central nervous system due to a slow virus which was transmitted by ritual cannibalism. It is found in the Fore people and some neighbouring tribes who live in the eastern highlands of Papua New Guinea. It is known to have been present for over half a century but was only delineated when Gajdusek (1973) began his studies of these people in the 1950s. The condition occurs in children of both sexes and in women, but is rare in men. The disease is most prevalent among the Fore people and accounted at one time for about 90 per cent of deaths in adult women. During the last 20 years the incidence of Kuru has decreased and the geographical area in which the cases occurred has been contracting. The localized occurrence of Kuru initially suggested that the disease might be genetically determined. However, anthropological studies revealed that the Fore people and nearby tribes still practised endo-cannibalism, although this was more in the nature of a social or religious rite than as a regular food source. The custom was for dead kinsman to be butchered, cooked, and eaten as part of a mourning ceremony. This was carried out by the women and the flesh was rarely eaten by older boys or by men. Proof that Kuru is a slow virus transmitted by the cannibalistic practices came when chimpanzees who had been inoculated with material from the brain of patients dying of Kuru developed similar neurological lesions to those found in humans.

Food can also be contaminated during preparation. Traditional weaning foods can be bacteriologically contaminated. For instance, an outbreak of botulism in Kenyan nomads was due to a contaminant preparation of sour milk in a calabash, and diarrhoeal disease is often most common in infants at the time of weaning.

Agricultural practices bring humans and animals into closer contact. This increases the likelihood of transmission to humans of diseases normally found in animals (zoonoses). Brucellosis in Kenyan highlanders is attributable to their habits of sharing their huts with their cattle and goats, and subsisting mainly on goat's milk often infected with *Brucella melitensis*.

Urbanization, affluence, and modernization

The overall decline in communicable disease in the wealthier communities of the world started when living conditions began to improve. The combination of social and economic advance followed by medical preventive measures are generally considered the main reasons for the observed decline. However non-communicable diseases, including heart disease, diabetes, and some forms of cancer, have increased considerably in these countries.

Coronary heart disease (CHD) is often said to be a disease of affluence. This explanation is far too simplistic. In England and Wales CHD death-rates were more common in upper social class men until 1951. However, by 1961 the death-rates for men in classes I and II levelled off, whereas they continued to rise in working class men. In 1971 the standardized mortality rates for social classes I and II were 88 and 91 respectively compared with death-rates of 108 and 111 for classes IV and V. Marmot (1980) has argued that when a population reaches a certain level of economic development, CHD is no longer a disease limited to the more affluent members of that society.

Coronary heart disease has also been said to be associated with modern urban living and with a diet rich in saturated fat, with smoking, and with a sedentary lifestyle. Data from England and Wales show slightly higher CHD mortality in urban than rural areas but there is no effect of size of urban area. However, the picture is somewhat different elsewhere. In former Yugoslavia and Puerto Rico CHD incidence is twice as high in urban as in rural residents. These two areas are not as developed or urbanized as Northern Europe or America but more so than the developing countries of Asia and Africa.

Little or no information is available from developing countries for urban–rural differences but there is considerable information on blood pressure which is a major risk factor for CHD. In most populations, the mean blood pressure level in males and females rises progressively with increasing age. This general statement holds true for most communities in the tropics and elsewhere but there are communities in which blood pressure does not appear to rise with age and in which the problems of essential hypertension appear to be virtually non-existent.

These exceptional minorities are of considerable interest and include relatively small and isolated communities, such as those of New Guinea highlanders, Kalahari San, Pacific Islanders, Australian aborigines, and

East African nomads. It is difficult to decide whether these groups are otherwise normal, or whether they merely reflect the presence of such factors as chronic infection, parasitism or malnutrition which somehow prevent the usual rise of blood pressure. Critical to this problem is the question of whether these isolated communities are capable of developing higher blood pressures under changed environmental conditions.

From studies carried out on young nomadic warriors in Kenya entering the army, there is evidence that a changing environment produces significant elevations in systolic blood pressure over a period of only two years. In New Guinea, the rural highlanders have low levels of blood pressure which do not rise with age, whereas an urban wage-earning group in Port Moresby showed much higher blood pressures in the older age groups.

The problem with these studies is the assumption that the groups differ only in degree of urbanization. There could be other, including genetic, differences between the groups. One way of overcoming some of these difficulties is to use migrant studies. Prior and his colleagues (1974) undertook studies on the Tokelau islanders of Polynesia. Because of population pressure there has been continued migration of islanders to New Zealand. Surveys of Tokelauans living on the islands and the migrants in New Zealand show a steady rise of blood pressure with age (Fig. 1.2). The blood pressures of the migrants are, on average, significantly higher than the sedentes (non-migrants). These results are in keeping with previous findings which report an urban – rural difference. However, what was unique about the Tokelau migrant study was that Prior and his colleagues went on to undertake a prospective study where they measured the blood pressures of potential migrants while still resident on the islands and remeasured them after migration some six years later.

The results (Fig. 1.3) show that the sedentes and pre-migrants had similar levels of blood pressures. Six years later the blood pressure of migrants had risen to a greater extent than the blood pressure of the non-migrants. Older immigrants show a greater difference in blood pressure from sedentes than younger immigrants which suggests that there is a decline in adaptability with age.

The obvious questions which arise from these studies concern those features which the low pressure groups have in common and those factors which might possibly be concerned in producing an increase in their blood pressure levels. An obvious physical feature of all these low blood pressure groups is their lean body build and the virtual absence of obesity. In situations in which the blood pressure rise results from an environmental change there is usually an increase in body fatness. It is unclear whether the observed rise in blood pressure with increase in body bulk is directly due to the accumulation of adipose tissue but what evidence there is suggests that the relationship is neither direct nor simple.

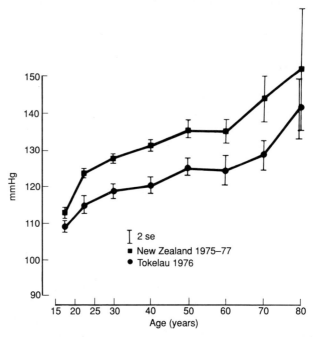

Fig. 1.2 Mean systolic blood pressure by age of Tokelauan men living in Tokelau and New Zealand. (I. A. M. Prior *et al.*, unpublished data.)

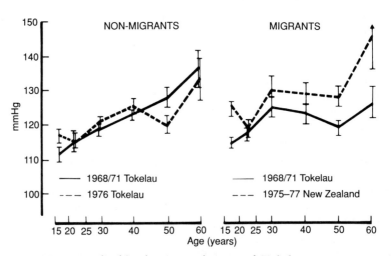

Fig. 1.3 Mean systolic blood pressure by age of Tokelauan men comparing non-migrants and men who subsequently migrated to New Zealand. (Prior *et al.*, unpublished data.)

There is considerable circumstantial evidence to suggest a positive correlation between the level of salt intake and the prevalence of hypertension
in a community. (Prevalence refers to the percentage of the population or
age group found to have a condition or to be infected at a given point in
time. Incidence is the number of new occurrences of a condition or infection
among uninfected children over a given period). In almost all the low blood
pressure groups salt intake is very low, whereas in those groups in which
a rise in blood pressure has occurred through social and environmental
change, salt intake has almost always increased substantially. Genetic
factors may play a part in determining which individuals are sensitive to salt
ingestion.

Another striking feature of the low blood pressure groups is their social
structure and its perpetuation by virtue of cultural or geographic isolation.
It is possible that the higher blood pressure represents a failure to adapt to
changing environmental conditions.

Marmot and others argue that the increase in risk of CHD occurs in situations of newly emerging urbanization or economic development. Populations at high risk are characterized by high mean cholesterol levels of
200 mg/ml or more, whereas populations at low risk have much lower
mean levels. High mean serum cholesterol is associated with high levels of
low density lipoproteins (LDL) and usually with low levels of high density
lipoprotein (HDL). HDL cholesterol appears to be a protective factor in the
development of coronary artery disease and levels are high in most populations with a low mortality. However, a few groups, such as the Masai of
Kenya and Tanzania, who have a very low incidence of CHD, have low
levels of HDL as well as low total serum cholesterol. In most African
populations serum HDL levels are high and tend not to fall with age as in
Western countries.

In addition to hypertension other risk factors include family history,
obesity, diabetes mellitus, cigarette smoking, physical activity, and per-
sonality type. However, it would appear that their effects are only
manifested in susceptible populations characterized by high mean serum
cholesterol levels. Thus, in low-risk populations, such as rural black
Africans, the addition of even multiple risk factors in individuals rarely
leads to the development of CHD. Atheroma even in the obese hypertensive
African patient with diabetes mellitus remains less common than in Western
populations.

Countries with a high rate of CHD have a national diet that is rich in
saturated fat. The reverse is not necessarily true. There are some populations, including the Masai and the Inuit, who have a high consumption of
fat but a low incidence of CHD. Both groups have low levels of serum
cholesterol. It seems probable that there has been a long-term biological
adjustment to their particular diets. The Masai have developed an efficient

feedback mechanism which suppresses their endogenous cholesterol synthesis. The low level of plasma cholesterol in Inuit, despite the high intake of animal fat, might be due to the large amount of polyunsaturated fatty acids in the tissues of the animals that provide their food.

Although saturated fat appears to be the best indicator of risk in a population, there is evidence that serum cholesterol levels may be influenced by other dietary factors. Burkitt and others suggest too little dietary fibre is an important risk factor. This has led to the notion of 'the prudent diet' which aims to reduce the level of saturated fat by altering the saturated to polyunsaturated ratio, and to increase the consumption of unrefined carbohydrate containing a high fibre content at the expense of refined carbohydrates. There is strong, although not conclusive, evidence that adoption of these measures along with secondary risk factors have led to the decline in mortality of CHD in the United States and Australia.

Beliefs

In many parts of the world, sick people first consult a traditional healer, and only later, if he or she cannot help, do they consult a medical doctor or visit a hospital. Ideas about the cause of disease influence whom people choose to help them. In Kenya, pulmonary tuberculosis among the Kamba people and infertility among women are attributed to spiritual causes so that the spiritual healer is consulted first and with most trust by the patient. Other beliefs include witchcraft, the hand of God, and 'something within'.

Contact with religious objects may enhance the spread of disease (Reynolds and Tanner 1983). The sharing of the communion cup practised by some Christian denominations may increase the risk of transfer of bacterial infections (Gregory *et al.* 1967). The followers of Islam ritually wash before prayers. In rural areas of Yemen and Bangladesh, where the water comes from storage tanks, those who wash run the risk of picking up infections from Guinea worm (Underwood and Underwood 1980), schistosomiasis (Farooq and Mallah 1966) or any other disease present in the water. Cholera spreads when water contaminated by infected excreta is ingested. In 1863, pilgrims brought cholera from India to Mecca and 30 000 pilgrims are thought to have died.

It has often been assumed that there is a close connection between the religious ritual of male circumcision and the absence of cervical carcinoma in women and penile carcinoma in men. Although Jewish women have low rates of cervical carcinoma relative to non-Jews in the United States (Weiner *et al.* 1951) there is little difference in cervical cancer rates in Lebanese Christians (uncircumcised) and Muslims (circumcised). Among the uncircumcised Amish the risk of cervical cancer was considerably less than among non-Amish controls.

Several studies have shown that circumcision is not a clear category as the uncircumcised may have short foreskins, which is a natural occurrence in about 20 per cent of all men. Furthermore, data on circumcision can be inaccurate; 33 per cent of wives incorrectly said their husbands were circumcised, whereas for men 7 per cent gave an incorrect response.

The consanguineous marriages of Muslims will also be expected to lead to increased amounts of homozygosity in their inbred offspring which might have biological repercussions if disadvantageous genes are present. Some studies of consanguineous matings have been found to show increased levels of congenital malformations, still births, and prenatal deaths, although this is not always the case.

Occupation

There are many examples of increased risk of disease associated with a specific occupation. Industrial lung diseases arise from both mineral and organic substances. Pneumoconiosis is associated with coal mining, gold mining can cause silicosis, and cocoa bean handlers may suffer from wheezing dyspnoea.

It is now known that asbestos is a causal factor in malignant mesothelioma and lung cancer and probably in cancer of several other sites. A consensus with regard to this relationship was reached only about 1960, despite the fact that some of the earliest observations were made over 50 years ago.

The earliest recorded observation on a relationship between asbestos exposure and lung cancer was made by Wood and Gloyne in 1934. They reported two cases of lung carcinoma seen at autopsy in 53 cases of asbestosis. Within the next four years similar case reports appeared in the literature from both the United States and Germany. The Germans quickly arrived at the conclusion that there was a cause and effect relationship, and in 1943 the German government issued a decree declaring cancer of the lung, when associated with asbestosis of any degree, an occupational disease. The combination of World War II and the unpopularity of German literature and laws after the war meant that it received little attention.

It was not until 1955 in England that Richard Doll carried out what was probably the first epidemiological study of asbestos exposure and cancer. He demonstrated that the incidence of lung cancer in exposed subjects was considerably higher than the incidence in the general population. Unfortunately, his small sample and the absence of any positive experimental data left many people sceptical. The turning point came in 1964 when Mancuso and Coulter reported a positive epidemiological study of workers in an asbestos textile plant and Selikoff, Chung, and Hammond (1964) showed a large excess of respiratory cancer among asbestos insulation workers.

The quantification of risk to human health is also important. A follow-up study of retired asbestos workers revealed that there was a linear association between the amount of dust workers had been exposed to by retirement and their subsequent mortality experience from cancer of the lung. The linear regression line indicates that one year's exposure to 1 million particles per cubic foot of asbestos dust, 8 hours a day, 5 days a week, will cause about 1 half-point increase in the standardized mortality ratio for lung cancer.

Sexual practices

The sexually transmitted diseases are common all over the world. In industrial countries they are a major medical and social problem and their incidence is steadily rising. Rates in Africa are alarmingly high. For example, in Ibadan, Nigeria, the prevalence of gonorrhoea among asymptomatic women was about 5 per cent, whereas among prostitutes and female hospital patients it was 15–20 per cent. In Kampala, Uganda, the prevalence is 10 000 per 100 000 and in Nairobi, Kenya, it is 7000. In Greater London the corresponding figure is 310 and in Atlanta, Georgia it is 2510. These are urban examples but the figures from rural areas also suggest elevated levels.

There are a number of factors which are responsible for the high rates in Africa. Low priority is given to control by health authorities so that there are inadequate facilities for diagnosis, treatment, and tracing of contacts. Self-medication is frequently practised which is quite often inadequate and this can result in the breeding of resistant organisms (e.g. penicillin insensitivity of gonococcus). The other major factor is caused by rural–urban migration. More and more young people leave their rural homelands and migrate to urban areas in search of work. A large number fail to find a job and the men often become vagrants and the women prostitutes. The rapid spread of gonorrhoea has been aided by its short incubation period and because it is usually asymptomatic in females. As travel increases and rural areas are developed the spread of the disease to these rural areas becomes inevitable.

Venereal syphilis is increasing in many industrialized countries. Syphilis is one of the treponematoses which are diseases caused by spiral organisms called spirochaetes. The treponematoses include venereal and endemic syphilis, yaws, and pinta. All these diseases are immunologically related and offer some cross-immunity against each other.

It is unclear whether the prevalence of syphilis is increasing in tropical Africa. One opinion is that because the incidence is increasing in industrialized countries then the same trend will occur in Africa. This view is strengthened by the fact that the eradication of yaws in the 1950s meant that a large number of people have become susceptible to venereal syphilis because of the removal of cross-immunity provided by yaws. Epidemics of venereal syphilis have been reported in areas where yaws was previously endemic. Very high rates of infection have been reported in Uganda, Ethiopia, and Nigeria. In early infectious syphilis the prevalence rates are 1000 for Kampala, 500 for Nairobi, 9 for Greater London, and 90 for Atlanta.

The other view is that there has been a steady decline in the incidence of syphilis in Africa over the last two decades aided by the increasing but indiscriminate use of penicillin and other antibiotics. This view is supported by the fall in number of patients seen with syphilis in urban hospitals. Whichever viewpoint is correct the incidence of syphilis is still considerably greater than in the industrialized countries.

By far the greatest challenge by disease facing the world is the spread of AIDS (acquired immune deficiency syndrome) caused by human immuno-deficiency virus (HIV). The HIV virus is very fragile and outside of blood or semen will die. It has a relatively long incubation period. AIDS must be distinguished from seropositivity or detection of antibodies to HIV virus in serum, although the majority of seropositives eventually manifest clinical expression. Currently it is estimated that between 5 and 10 million people are infected with HIV. Although the majority of infected are adults there are reports of dramatically increased numbers of HIV-infected children.

There are three primary modes of transmission of the virus: (i) sexual activity; (ii) blood and blood products; and (iii) perinatal transmission. Sattenspiel and Castillo-Chavez (1990) describe the seven high-risk groups as: (1) homosexuals; (2) intravenous drug users; (3) sexual contacts of intra-venous drug users (both male and female); (4) recipients of blood trans-fusions (in the United States prior to 1985); (5) haemophiliacs (because of transfusions with affected blood); (6) sexual contacts of transfusion recipients and haemophiliacs; and (7) children of infected individuals.

In the United States, the proportions in each of the high-risk groups reported with AIDS has remained relatively constant during the 1980s with 65 per cent of all cases in homosexuals, 20 per cent among intravenous drug users, and the rest in the other high-risk groups. However, these propor-tions are expected to change and recent reports show a five-fold increase in the number of homosexual and bisexual men practising celibacy in the United States.

Considerable efforts are being made to provide realistic models for the spread of HIV in the United States and elsewhere. However, there is limited information on the nature of sexual behaviour, the preventive measures taken, age distribution of the population, and behaviour associated with intravenous drug users including use of shared needles. Other factors which need to be taken into account are the pattern of infectivity and the pro-gression of the disease within an individual. Consideration of all these factors makes a variety of outcomes possible depending on particular choices of parameters.

Valleroy and her colleagues have modelled the course of the AIDS epidemic in East African women aged between 25 and 35 years of age using actual figures for infection from pregnant women in five cities in East Africa. The incidence of the infection ranged from 0.4 per cent in Maputo, Mozambique, to 24.1 per cent in Kampala. Valleroy et al. calculated how many extra women would die before their children reached 5 years of age because of AIDS. The increase begins to show by 1992. They then estimated how many extra children would be orphaned because of these deaths, adjusting the figures to take into account that some children would die from AIDS before the age of 5 years.

The model predicts that in Bujumbara in Burundi and Kampala, twice as many women in the 25–35 year age group who will die in 1992 will have succumbed to AIDS. The number of children orphaned in the process will double. Running the model by the year 2015 for the whole of sub-Saharan Africa produced a figure of 2.4 million deaths from AIDS that year alone with 78 per cent of the women of childbearing age dying from AIDS. This could leave 16 million children orphaned in sub-Saharan Africa.

Such a huge mortality will not only devastate the labour force and overload an already stretched health service. It will also leave countries with the enormous burden of caring for millions of orphans. Already there are signs of the turmoil to come. In the areas hit hardest by the epidemic, such as southern Uganda and northern Tanzania, there are grandparents who have been left to care for 14 or so children. About 30 to 40 per cent of children born to infected mothers also carry HIV. Between 50 and 90 per cent of these children die from AIDS before they reach 5 years old.

The problem of children orphaned by AIDS is not unique to East Africa. Parts of the Caribbean and New York City will soon have to cope with large numbers of parentless children. In Port au Prince, Haiti, the number of orphans will increase by 60 per cent by 1992.

As yet there is no vaccine ready to be tested for its ability to protect against HIV infection. Nevertheless, several candidate vaccines are now being tested using healthy volunteers to evaluate their safety and the kind of immune response they produce.

Agricultural practices

Infections associated with agricultural practices have already been mentioned. Sources of infection are either animals (e.g. sleeping sickness, bovine tuberculosis), soil (e.g. hookworm), water (e.g. Guinea worm) or brought about by dietary deficiency and trauma.

Schistosomiasis

One disease which has dramatically increased in prevalence as a result of agricultural practices is schistosomiasis. This disease, also called bilharziasis, is a parasitic disease which affects over 200 million people and poses a threat to 400 million more in at least 76 countries. The disease is continuing to spread despite technical advances and refinements. Schistosomiasis, which is rarely lethal, is caused by blood flukes and it rivals malaria and sleeping sickness in being the greatest threat to human health in tropical Africa.

The fluke, a comparatively large flatworm was described as early as 1851 by a German named Theodor Bilharz after he noticed one during an autopsy carried out in Cairo. It is a bisexual organism with the bulky male parasite providing protection for the flimsier female worm whose main function is reproductive. The female is about two centimetres in length and the male about half as long. The disease was originally named after Bilharz (by the entomologist Spencer Cobbold) but purists subsequently claimed that the worm had already been called schistosoma. Both synonyms are in use although by the rules of biological nomenclature priority is given to schistosoma and schistosomiasis. In Africa the disease also goes under the name of 'Bill Harry' or 'snail fever', which provides an indication of the parasite's intermediate host.

Different species of the fluke cause three main forms of the disease. Two occur in Africa (*Schistosoma mansoni* and *S. haematobium*) and the third in the Far East (*S. japonicum*). The three species have similar life cycles and develop over a succession of stages—egg, miracidium, sporocyst, cercaria, schistosomulum, and adult.

The male and female forms which cause schistosomiasis in the urinary form (*S. haematobium*) live together in the veins draining the human bladder (Fig. 1.4). In them they mate periodically. The female parasite lays hundreds of eggs which migrate through the bladder wall and are discharged in the urine. At least 6 days elapse between oviposition and egg excretion. The penetration of the bladder causes bleeding and the first sign of the infection is the passage of bloody urine. Later on, the disease may be complicated by the formation of stones and polyps in the bladder, and the latter sometimes become cancerous.

A large proportion of the eggs fail to work their way through the bladder wall and remain unhatched in the body. These errant eggs are washed back into the host's bloodstream and are scattered throughout the patient's system where they presently die. The piling up of dead eggs blocks the normal flow of body fluid and they act too as irritating foreign bodies which results in localized formation of fibrous tissue as the host attempts to wall them off. The fibrosis in turn leads to all manner of complications. Thus,

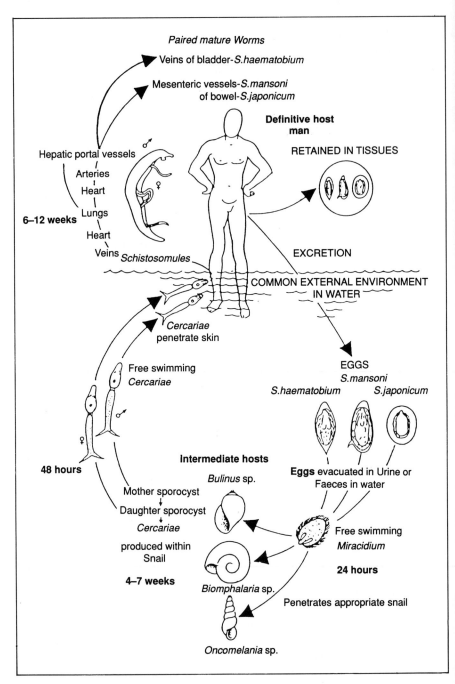

Fig. 1.4 The life cycle of the *Schistosome* parasite.

ova which lodge in the brain often cause epilepsy. Many lodge in the liver which develops cirrhosis.

For the life cycle to continue the egg must pass into fresh water where the protective shell ruptures and free-swimming larvae called miracidia are released. Miracidia swim actively (2 mm/s) and they seek out a particular fresh water snail. Miracidia can infect snails located 33 cm deep and 86 cm away from their starting position and they remain infective for about 8 to 12 hours. A large number of factors influence the infection of snails by miracidia and they include temperature, water turbulence, flow, ultraviolet light, as well as the age of the snail and miracidia and their numbers.

Penetration of the snail intermediate host occurs in 70 per cent of instances via the foot. After penetration the larva disappears and within a few days develops into a mother sporocyst. This only takes place if the appropriate species of snail has been entered otherwise the larva is destroyed. In the snail production of cercariae which are all of the same sex occurs and this takes 4–5 weeks for *S. mansoni*, 5–6 weeks for *S. haematobium*, and over 7 weeks for *S. japonicum*.

The cercariae leave the snail following a daily rhythm which is unique to each species of schistosome. The daily output is enormous and in laboratory conditions Barbosa *et al.* (1954) counted the daily average as 4598 with a maximum of 17 600. Over an 18 week period the total cercariae production was 682 360. However, in natural conditions it is thought that production rarely exceeds 1500 cercariae/day for *S. mansoni* and 2000/day for *S. haematobium*.

Cercariae do not feed and their principle energy reserve is glycogen stored in both the body and tail. The size of the reserve significantly influences longevity and the half-life is about 30 h at 15 °C, 16 h at 25 °C, and only 8 h at 35 °C. The act of penetrating a human host uses between 22 and 35 per cent of the glycogen stores.

How cercariae locate the human host is still poorly understood. The environmental stimuli to which cercariae respond by changing their activity pattern are precisely those which a human wading through water generates, namely turbulence, shadows, and the presence of human skin substances. After penetration of the host skin the larva changes to become a schistosomulum which has a tailless worm-like appearance. The passage of the schistosomulum through the subcutaneous tissues takes about 48 hours after which it is transported to the right heart and lungs. Growth occurs in the lungs. The schistosomulum then passes to the hepatic portal vein but the route is uncertain. Most sexually mature worms leave the liver when they have mated and migrate to either the veins of the bladder (*S. haematobium*) or to the mesenteric veins (*S. mansoni* and *S. japonicum*).

Adult worms can live for 20 to 30 years but the average is between 3 and 8 years. As is the case with most helminths, the adult schistosomes do not multiply in the definitive host. This leads to a situation in which most individuals carry very few parasites and only a small proportion are heavily parasitized.

The bilharzia flukes boast a long ancestry. They infected the ancient Egyptians, and typical eggs have been recovered from mummies dating back to 1500 BC. It is supposed that wars and migrations carried the flukes from the Nile valley to the central African lakes whence this inheritance from Egypt became distributed throughout most of the continent.

Other aspects of human behaviour are responsible for the continuing spread of the disease. Europe first learned about schistosomiasis following Napoleon Bonaparte's expedition into Egypt in 1798. His soldiers soon began to suffer severely from a disease characterized by the painful passing of bloody urine. A century later schistosomiasis became such a problem to British troops serving against the Boers that for years the British government annually paid out £10 000 to soldiers incapacitated by the disease.

The bilharzia fluke appears to have a talent for harassing the military. When the Americans stormed ashore on Leyte in October 1944 they encountered these same parasites, about which they had received no warning. It was not until New Year's Day 1945 that the first military cases were diagnosed among patients in an evacuation hospital. In the end, 1700 men were put out of action by the fluke at a cost of 300 000 fighting man days and US$3 million. The lesson of the dangers from bilharzia had not been learned five years later when 50 000 Chinese communist soldiers assembled during 1950 for the projected invasion of Formosa. Schistosomiasis became widespread among them and led to the abandonment of the campaign and the survival of Chiang Kai-shek's island nation of Taiwan.

No account of the discovery of the fluke's cycle and early treatment would be complete without some mention of the heroic work of an American doctor called Barlow. To prove the method of schistosomal transmission to man, Dr C. Barlow placed 224 live cercariae on his belly and waited patiently for a succession of nettle-like stings to announce the entry of the schistosomal larvae into his skin. He presently went down with a severe bilharzial infection which demanded so many injections of tartar emetic that, in his old age, Barlow wrote 'even today I shudder every time I see a hypodermic needle'.

Control measures

In 1984, the World Health Organization announced that schistosomiasis is caused by people not snails. The acceptance of this idea, has and is, causing a revolution in the concept of schistosomiasis control.

Control involves interruption or reduction in the transmission of the parasite by either halting the introduction (or re-introduction) of infection into the community or by prevention of exposure to cercarial-infected water. Therefore, the aims of control programmes focus on trying to reduce infection in water by the use of focal mollusciciding to kill infected snails; reducing contamination by chemotherapy and provision of latrines; reducing exposure by improvement in water supplies; reduction in water contact by building fences and bridges; and, finally, educating people about the disease and how it can be prevented. Even so there has been concern over the recurring introduction of chemicals into fresh water and the rising costs of chemotherapy.

One project which has introduced a comprehensive strategy for the control of schistosomiasis is The Blue Nile Health Project of Sudan. Sudan is the largest country in Africa but much of it is desert or semi-arid flat plains. The country's major asset is the Nile, and since 1924 with the completion of the Sennar Dam, improved water management has led to increasing use of fertile Nile flood plains. The famous Gezira scheme was opened in 1925 with 250 000 acres of land irrigated via a network of open canals. Today the Gezira-Managil scheme totals 2 million acres and the cotton it produces generates over 50 per cent of total export earnings.

However, the price paid for agricultural development has been a dramatic increase in both the prevalence and intensity of schistosomiasis. (Intensity of infection relates to the worm burden and it is indirectly assessed by egg output). Over half of the residents of these agricultural areas were infected and in school children the prevalence was over 80 per cent.

In the early 1980s new anti-schistosomal drugs were being developed and praziquantel became the first drug considered safe to use on a large scale. Praziquantel was used for mass treatment in villages with prevalence rates of more than 40 per cent; in villages where the prevalence rate was lower only infected persons were treated, i.e. selective population chemotherapy. Further measures taken included focal mollusciciding, weed removal in some areas, building of pit latrines, provision of safe water supplies, and the introduction of health education programmes. The overall prevalence dropped from about 50 per cent to 11 per cent in one year.

In addition to the Gezira-Managil, a new irrigation area, the Rahad scheme, opened in 1974 and the indigenous population was initially free of schistosomiasis. However, a large number of immigrants had been attracted to the area, many of them from endemic areas or with schistosomiasis contracted in the Gezira. The prevalence of schistosomiasis is rising and in 1988 had reached nearly 20 per cent.

Like many diseases, schistosomiasis is largely caused by human behaviour—in this case, principally water use practices and indiscriminate urination and defaecation, but also failure to take advantage of available screening practices or to comply with medical treatment. However, it is important to avoid blaming people and their behaviour exclusively, especially when appropriate control measures are not widely available.

This brief description of the life history and ecology of schistosomiasis illustrates the importance of agricultural practice, war, migration, human behaviour, customs, and taboos in the spread of infectious disease.

Other diseases

The prevalence of many other diseases is increasing. Stoll (1947) in his classic review: 'This wormy world' estimated the world prevalence of hookworm to be some 457 million cases. Later he revised his estimate to about 600 million (Stoll 1962) and the latest calculations put the figure at between 800 and 1500 million cases with about 55 000 to 60 000 deaths a year from hookworm disease (WHO 1984b; Schofield 1985; Bruyning 1985). Indeed, roundworm, whipworm, and hookworm are among the 10 most common infections existing in the world. Although morbidity and direct mortality are relatively low in proportion to prevalence, these intestinal helminths can produce adverse effects on nutritional and immune status as well affecting cognitive performance (Kvalsvig 1988; Halloran *et al.* 1989). Morbidity in young workers, the most productive sector of the population, can also be economically important.

In the tropical and sub-tropical zones, against the general background of over-population, deficient or absent sanitation, limited access to potable water supplies, environmental conditions favouring helminth transmission, and widespread socio-economic deprivation, helminth infections have become increasingly important even though the principles of surveillance and control are well understood.

The common large roundworm (*Ascaris lumbricoides*) is the most cosmopolitan of all helminths. Females living in the human small intestine lay fertilized eggs which are excreted in faeces. When deposited in moist, loose, shady soil the eggs develop into larvae after about 2 to 3 weeks incubation. If swallowed by humans, usually via fingers contaminated by polluted soil, via food or drink, especially unwashed vegetables or fruit, or from young children eating dirt, the larvae emerge from the eggs in the small intestine, penetrate the intestinal mucosa and are carried in the portal circulation through the liver to the lungs. After a period in which moulting occurs, the larvae migrate through the pulmonary capillaries to the alveoli and then via the bronchioles, bronchii, and trachea to the oesophagus. During this phase considerable growth occurs. After being swallowed, a final moult and growth to maturity takes place in the small intestine. The time from egg ingestion to egg laying by adult females is about 60 to 80 days. The life span of the adult worms is 9 to 12 months and humans are the only important natural host of this species.

Ascariasis has a world-wide distribution and the prevalence is between 50 and 75 per cent or more in many African and Latin American countries. Neither sex nor ethnic origin confers any protection from infection and differences between groups can usually be related to social customs or work habits. There are quite often age differences in prevalence, with higher prevalence in young children, but these result from toddlers crawling and playing on faecally contamined earth. In general, soil-transmitted intestinal infections are less frequent and of lesser intensity in urban than in rural areas.

All nematodes are prolific egg layers and a single female *Ascaris* deposits about 200 000 eggs per day (Faust and Russell 1964) and worm burdens in excess of 100 are not uncommon. The global population of *Ascaris* worms has been estimated as 7800 million and because each female adult has a reproductive potential of over 200 000 eggs per day, the daily global contamination of *Ascaris* eggs is some 10^{18}. Further realization of the problem can be appreciated from the 1975 figures which estimated that in the rural areas of developing countries, excluding China, 1190 million people (85 per cent of the total population) lacked adequate sanitation. The degree of contamination of the environment is enormous and is solely due to inadequate faecal disposal.

Control of the soil-transmitted helminthiasis has been less successful than many other campaigns of preventive medicine. The combined effects of treatment and the introduction of adequate sanitation in the earlier years of this century, reinforced by health legislation, widespread education, and rising living standards, diminished the prevalence of hookworm and other intestinal helminths in Europe and North America until they have become of minor importance.

One of the earliest organized campaigns against hookworm was the extensive programme conducted in Germany from 1903 to 1914 which aimed at reducing the prevalence of infection in mining communities. The programme focused on the treatment of infected miners, the exclusion from underground work of new employees discovered to have hookworm infection, and the installation of sanitation. The success of the programme was instrumental in promoting control efforts in other European countries. In the United States, the Rockefeller Sanitary Commission was established in 1909 to combat hookworm disease. Since 1913 numerous campaigns have been conducted against hookworm and the areas of activity include the West Indies, Central and South America, Sri Lanka, India, Thailand, New Guinea, and Australia.

Control of helminths involves education, sanitary disposal of human faeces, and the use of chemotherapy. In Bangladesh, the prevalence of

Ascaris, *Trichuris*, and hookworm is high and some non-government organizations have started to deworm children regularly. However, without the other control measures the maximum benefit will not accrue.

The Blue Nile Health Project is a control programme and must not be confused with eradication which is the complete and permanent cessation of transmission. With the present methods eradication will rarely be achieved for schistosomiasis. Less ambitious programmes, however, may achieve reduction in the level of transmission, in the clinical consequences of infection, and may ameliorate the socio-economic sequelae of schistosomiasis.

Malaria

Eradication campaigns against malaria have produced variable results. In developed countries and particularly where malaria was epidemiologically unstable, spectacular successes have been achieved. In many developing countries, however, results have tended to be unsatisfactory.

Malaria is a disease caused by protozoa of the genus *Plasmodium*. Nearly 120 species of *Plasmodium* exist of which 22 species are found in primates. Only four species of *Plasmodium*, *P. malariae*, *P. vivax*, *P. falciparum*, and *P. ovale*, parasitize humans. In the past, infections caused by these four species were given colloquial names thus *P. vivax* was known as benign tertian, simple tertian, or tertian; *P. malariae* as quartan; *P. falciparum* as malignant tertian, subtertian, tropical, or pernicious, and *P. ovale* as ovale tertian. These terms are used less frequently and malariologists increasingly refer to falciparum malaria, etc.

Malaria is characterized by fever, sometimes recurring every second or third day, anaemia, splenomegaly, headaches, and a wide variety of other symptoms. Cerebral malaria, caused by *P. falciparum*, leads to unconsciousness and death. It is estimated that at least 1 million Africans die each year from malaria, mostly children under 6 years of age. In the past, malaria was much more widely distributed and was present in most of Europe.

The life cycle of malaria is essentially the same in all species (Fig. 1.5). It consists of an exogenous sexual phase (sporogony) with multiplication in certain *Anopheles* mosquitoes and an endogenous phase (schizogony) with multiplication in the vertebrate host. The endogenous phase includes the development cycle in the red blood cell (erythrocytic schizogony) and the cycle in the parenchymal cells of the liver (exo-erythrocytic schizogony). Zoologists distinguish between the definitive and alternative host. The mosquito is the definitive host because this is where sexual development of the *Plasmodium* takes place. Increasingly, these terms are being replaced by vertebrate and invertebrate hosts.

In an individual in the absence of re-infection, *P. falciparum* infections tend to die out without treatment in 18 months to 2 years, *P. vivax* and

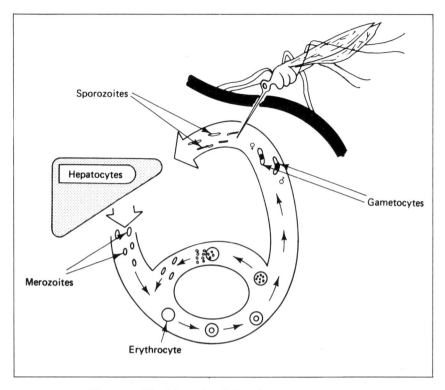

Fig. 1.5 The life cycle of the *Plasmodium* parasite.

P. ovale in 3 to 4 years, *P. malariae* may die out in 3 to 4 years but may persist for 50 years.

From the point of view of human malaria the most important demographic feature of the vectors is their death-rate, in particular whether the mosquitoes are likely to die before the parasites they have ingested have had time to complete their cycle and be transmitted to a new host. At least under tropical conditions few mosquitoes die of old age, and the chance that one will survive for n days is therefore the nth power of its chance of surviving for one day (p). If it has survived for n days a mosquito then has a further expectation of life of $1 / -\log_e p$ days.

Desiccation is an important cause of death in mosquitoes, and their chances of survival are therefore much lower in the dry season tropical grasslands than in humid equatorial regions. Table 1.4 shows the effect of mosquito mortality on malarial transmission. The high mortality example ($p = 0.78$) is *Anopheles culicifacies* of southern India, but could be matched in most drier parts of the tropics. The example of low daily

TABLE 1.4. The effect of mosquito mortality on malarial transmission

		High mortality ($p = 0.78$)	Low mortality ($p = 0.95$)
A.	Mosquito bites an infective human		
B.	Chances that mosquito will survive 12 days (time required for P. *falciparum* cycle in a mosquito at 24°C) $= p^{12}$	0.05	0.54
C.	Mosquito's subsequent expectation of life $= 1/-\log_e p$ days	4 days	20 days
D.	Number of humans to which the mosquito, which bites man every other day, will transmit parasites during this period $= (C)/2$	2	10
E.	Potential number of infections in man from a single infected mosquito $= (B) \times (D)$	0.1	5.4

mortality ($p = 0.95$) is typical of A. *gambiae* in the more humid parts of west and central Africa. A decrease in daily mortality from 22 per cent to 5 per cent thus increases the potential number of new malaria cases by a factor of 50, and shows how mosquito mortality is probably the most important single factor determining whether falciparum malaria will be seasonal, epidemic, and unstable (as in most of its range) or highly endemic and stable (as in some parts of equatorial Africa and South East Asia).

It takes many years of constant re-infection for the human host to become more or less immune to falciparum malaria. However, soon after the first infection immunity begins to limit the numbers of circulating gametocytes, the only form of the parasite which is infective to the mosquito. When malaria transmission is intense throughout the year (as in the low mosquito mortality example shown in Table 1.5), the level of population immunity rises to the point where only young children remain infective to mosquitoes. But where transmission is seasonal, for example in tropical savannahs with a well-marked dry season, the whole population can become infective during the rainy season.

The mosquito and human immunity factors can be combined in a crude formula to give the Basic Reproduction Rate of the malaria parasite (Table 1.5). In example A, a high mortality mosquito transmits malaria in a virgin population and in example B, a low mortality mosquito transmits malaria

TABLE 1.5. Basic Reproduction Rate of the malaria parasite

	No. of days a human remains infective	$\times$	No. of mosquitoes / No. of humans	$\times$	No. of times a mosquito bites man daily	$\times$	Prop. of mosqs. surviving 12 days	$\times$	Their further expectation of life	$\times$	No. of times a mosquito bites man daily	$\times$	Prop. of these bites which are infective to man	=	Basic Production Rate
Example (A)	80		20		0.5		0.05		4		0.5		0.1		8
Example (B)	4		10		0.5		0.54		20		0.5		0.01		1.08

in a highly immune population. Situation A could be met on tropical grasslands of Asia, Africa, or the New World and is found with seasonal or epidemic malaria which lays low whole communities. In example B (which is not mathematically rigorous) the malaria parasite reproduction rate is brought down by acquired immunity which reduces the infectivity of man for mosquito (first factor) and of mosquito for man (last factor). However, the low mosquito mortality (middle terms) ensures continued transmission, often throughout the year. As Macdonald (1973) showed, in endemic malaria an equilibrium is finally reached in which the net reproductive rate of the parasite is a simple function of the mosquitoes' expectation of life and infectivity to man. In this situation adult humans are highly immune and clinical malaria may be restricted to the younger children. There is, therefore, less disruption of economic and social life and observers can gain the false impression that malaria is not an important disease in the community.

In situation A where vector mortality is already high, mosquito control by insecticides is usually effective in eliminating malaria. By this means, with only a minor role played by chemotherapy, malaria was rapidly eradicated from the whole of Mediterranean Europe and many parts of Asia, America, and the Pacific.

In situation B, the low performance of the parasite where population immunity is high might suggest that control measures could easily push the reproduction rate below 1 and therefore eliminate malaria. Unfortunately this is not so because any further lowering of transmission reduces population immunity, which increases the reservoir of humans infective to mosquitoes, and this in turn restores transmission to a higher level. In other words, the disease is stable.

The great stability of malaria where mosquitoes are long-lived and bite humans regularly is a major problem in the eradication of the disease, particularly in Equatorial Africa and some parts of South East Asia. Very few mosquitoes are needed to maintain transmission (control by DDT or other insecticides has to be very thorough) and very few human hosts are needed to provide an effective reservoir of infection (mass treatment by chemotherapy must be equally thorough).

For the future it appears that the development of a malaria vaccine is likely to provide the most promising form of preventing the disease.

SECTION 3: THE BIOLOGICAL EFFECTS OF DISEASE

Disease and mortality

Infectious and parasitic diseases are a major cause of morbidity and mortality in most developing countries. More than 30 per cent of deaths in children in their first five years of life are due to diarrhoeas, resulting in as many as three to five million deaths annually. Respiratory infections are a cause of high morbidity in the developed world, but in the Third World these infections, primarily pneumonias, are another major killer of children, with a world-wide estimate of 2.2 million per year. AIDS is a growing cause of mortality in developing and developed countries.

Parasitic diseases are endemic in most poverty stricken areas. Malaria continues to take its toll with some 150 million people affected annually and about 1 million children dying every year in tropical Africa alone. Livingstone (1971) has speculated that malaria may have killed more human beings than any other disease. Probably some 200 million people are infected with one or more of the lymphatic dwelling parasites and another 20 to 40 million suffer from onchocerciasis.

There is no doubt that the poorest countries suffer the most disadvantage. The poorest countries produce the greatest proportion of new births yet they have the greatest proportion of infant wastage and have the lowest parental literacy levels. In addition, the poorest have the worst nutrition and the least access to safe water. A World Bank survey in 1979 calculated infant mortality rates (death in the first year after birth) per 1000 liveborn to be between 49 and 237 for low income countries excluding India and China; 12 to 157 for middle income countries and 13 for industrialized countries. Child mortality (deaths per 1000, 1 to 4 years of age) were 18, 10, and 1 in the low, middle, and industrialized countries respectively.

The levels of maternal mortality and its causes are receiving increasing attention. Maternal death is defined as 'the death of a woman, while pregnant or within 42 days of termination of pregnancy, from any cause related to or aggravated by the pregnancy or its management but not from accidental or incidental causes' (WHO 1979). Data on the incidence of maternal mortality world-wide are patchy and before 1900 scanty. In the sixteenth to eighteenth centuries, mortality among members of the ruling houses of Europe was about 2000 per 100 000 live births. During the first half of the nineteenth century the level was about 1500 per 100 000. Subsequently, it fell to 549 (1891–95) and then to 394 (1911–13).

In developed countries nowadays there are between 5 and 30 maternal deaths per 100 000 live births and these compare with 450 per 100 000 live births in developing countries. In 1986, the World Health Organization calculated that 86 per cent of the world's births and 99 per cent of maternal deaths take place in developing countries. The average risk of a woman in Africa dying from maternal causes is 1 in 20 compared to 1 in several thousand for a European woman. The principal factors affecting maternal mortality in developing countries include obstetric complications, socio-economic, age, parity, birth interval, availability of care services, and reproductive behaviour.

Among pathogenic causes, haemorrhage accounted for 27.8 per cent of all maternal deaths, followed by illegal abortion (18.6 per cent), toxaemia of pregnancy (16.8 per cent), sepsis and obstructed labour (11.4 per cent), and ruptured uterus (5.8 per cent). There is a U-shaped relationship between maternal death and age with disproportionate risk of death among women in early and late reproductive years. In Bangladesh, relative risk of maternal death for women under 20 years of age is 1.8, or 80 per cent more than women in their twenties. For women over 35 years of age, the risk of death was more than twice that of women in the 20 to 35 year age group of the same parities. In Bangladesh, 15 per cent of all pregnant women were aged over 35 years.

High parity is common in developing countries. The relationship between maternal death and parity is U-shaped with nulliparous and multiparous (over 7) being at higher risk of death. The risk of maternal death to older pregnant women of parity 7 or more is very high. The risk of death is also higher for mothers whose pregnancies are close together. The lowest risk is for women whose delivery was 3–4 years after the previous live birth, with increasing risk for shorter intervals.

Maternal height is often used as an indirect indicator of pelvic size. Obstetric experience in many countries has proved its validity as a predictor of obstructed labour. One study in Sierra Leone found that women under 150 cms accounted for 85 per cent of such cases.

The association between previous obstetric complication and poor outcome of the next pregnancy is widely recognized. A study in Papua New Guinea found a very strong association between history of prolonged labour, operative delivery, third stage complications, as well as perinatal death and recurrence of complicated labour.

Anaemic women are at greater risk of maternal death, particularly from haemorrhage but also from sepsis and in extreme cases from heart failure. In the 1970s, anaemia was estimated to be an underlying factor in 40 to 50 per cent of maternal deaths in developing countries.

Maternal care services are also important. Kwast (1989) has estimated that of the 128 million births in the world each year, 58 million are not attended by a trained person. Without assistance of trained personnel, the mothers deliver 66 per cent of the infants in Africa, 36 per cent in Latin

America, and 47 per cent in South East Asia; 50 per cent of maternal deaths occur at home or en route to hospital.

Maternal mortality is exacerbated by low contraceptive use in many developing countries. Family planning can help to reduce maternal mortality by reducing the number of pregnancies or averting unwanted and high-risk pregnancies. Using the data of the World Fertility Survey it was estimated that the overall median proportion of maternal deaths which would be prevented is 29 per cent, based on currently married women aged between 15 and 49 years of age who said they did not want any more children but were not using an effective contraceptive method.

There are still taboos and misconceptions about the various methods of child spacing. In polygamous societies there is the additional complication that there may be competition between wives to have as many children as possible.

Disease and life expectancy

Historical events, changing farming and demographic systems, as well as evolution of pathogens and hosts, have all contributed to current disease distribution patterns. However, it is quite often difficult to make rigorous district, regional, national, and international comparisons simply because reliable information is not available. International comparisons of morbidity quite often reflect the efficacy of the disease reporting system or the willingness or otherwise of governments to publish information. In the past, information has often been suppressed concerning plague, typhus, and yellow fever; and only recently concerning epidemic and endemic famine.

Even if information is available difficulties remain. Comparisons of crude death rates in different countries are invalid because the age distributions vary between populations. A death-rate specific to a particular age group would be less vulnerable to false assumptions. The one most frequently used for comparisons is the infant mortality rate but this measure also suffers from some disadvantages mainly because it provides no information on other age groups. A low infant mortality rate might hide the fact that infants later die from malnutrition, measles or gastrointestinal infections. Consequently, the life expectancy at birth, which takes into account the specific mortality of each age group, is a better indicator of the chances of survival of the newborn.

Life expectancy at birth varies by region and with climate. The parallel lines on each side of the equator signify the Tropics of Cancer and Capricorn respectively (23° 27′ North and 23° 27′ South of the Equator) and it is common to refer to diseases occurring within this band as tropical diseases. The further outside the tropical latitudes the colder is the winter. The area subject to tropical-like infections is wider than the tropics because the principal insect vectors, the main agents of this extra disease load, can withstand all but freezing temperatures. The area free of severe winter frosts is limited by Australia and Brazil in the southern hemisphere, and Mexico and India in the northern hemisphere. This does not mean that the populations outside the frost-free region will be entirely free of vector-borne diseases. Flea- and louse-borne infections prosper in the coldest of climates and some malaria- and arbovirus-carrying mosquitoes can survive inside the frost zones. Many populations including those in North Africa, Pakistan, North India, Bangladesh, South Argentina, Chile, and Australia live outside the tropics but nevertheless are at high risk, even on a seasonal basis, from vector-borne disease.

It is suggested that the populations who live in the warm climate countries tend to have shorter lives than those in the temperate and cold climates. Why is this? Table 1.6 shows disease-specific death-rates for a number of common causes in 12 warm climate populations. As good data reporting is an essential prerequisite for being included the 12 countries are certainly not alone but may not be truly representative of all warm climate countries. But at least they provide a clue as to what is happening.

The first three columns of birth- and death-rates, and the rate of natural increase signify that all these are growing populations. The figures of the fourth column, the infant mortality rate have been used to order the countries from the most to the least favoured. It is evident that there is a rough association between birth-rate and infant mortality. It also appears from the disease-specific columns that the higher the infant mortality the more frequent are the deaths at all ages due to 'all infections and parasites' and specifically to the diarrhoeal diseases, tetanus, measles, and pneumonia. The eccentric data for tuberculosis suggest that many deaths from this disease are not correctly ascribed to it. Furthermore, in its pulmonary forms it causes death mainly in adults and thus will be less important in populations, such as those of Egypt and Mauritius, which comprise largely children and young people.

Deaths from accidents are seen to be appallingly frequent in most of the warm climate countries. As in Lagos, Nigeria road accidents have become the major cause of death in young adults in those countries in the process of economic development. In contrast, at least as far as the official notifications are concerned, deaths from tropical diseases occur less frequently than might be expected. Deaths ascribed to nutritional causes are also not especially frequent probably because the malnourished usually succumb from an infectious episode.

Table 1.7 shows the notifications to the World Health Organization of the occurrence of the principal diseases (frequency per 100 000 population) in 12 small or medium sized warm climate countries. In this table, the countries are arranged by incidence of tuberculosis. Such data need to be

TABLE 1.6. Birth-and-death data for 12 warm climate countries

	Fiji 80	Singapore 81	Australia 80	Thailand 80	Cuba 81	Kuwait 79	Mauritius 80	Sri Lanka 80	Ecuador 78	Egypt 81	Guatemala 80	Mali 76
Birth-rate per 1000	29.4	17.0	15.4	23.2	13.9	37.9	26.1	27.6	29.1	37.6	41.8	43.2
Death-rate per 1000	6.4	5.3	7.4	5.3	5.9	4.0	8.1	6.1	7.2	10.1	9.8	18.1
Natural increase (%)	23.0	11.3	8.1	17.9	8.0	33.9	18.0	21.5	21.9	27.5	32.0	25.1
Infant mortality rate/1000	9.9	10.8	10.7	13.3	18.5	31.1	32.2	37.1	64.4	73.5	85.9	120.9
Death-rates/100 000												
All infections + parasites	37.4	21.5	3.5	48.3	15.4	115.7	45.6	19.4	186.0	31.6	397.8	–
Diarrhoeal disease	17.8	2.3	0.6	11.7	5.5	65.2	31.3	49.7	106.2	7.5	190.7	–
Respiratory tuberculosis	4.7	12.3	0.4	14.6	2.8	15.2	2.6	10.5	14.7	5.4	14.1	–
Tetanus	0.2	0.2	–	1.9	0.2	2.6	1.8	–	9.9	6.7	2.6	–
Measles	0	0.1	0.1	0.2	0.5	3.2	–	0.5	25.6	2.6	105.6	–
Cancer	32.8	103.5	153.5	21.0	106.3	72.9	46.8	29.5	37.9	19.9	27.8	–
Mental + epilepsy	1.8	1.0	7.0	0.5	1.6	1.4	5.7	–	9.4	2.1	9.8	–
All cardiovascular	157.2	170.6	385.3	43.2	237.7	225.9	250.0	71.5	92.1	205.7	48.4	–
Pneumonia	28.4	40.4	13.9	10.1	41.4	–	28.3	–	49.5	45.1	96.3	–
Pregnancy + childbirth	3.0	0.7	0.3	4.0	1.9	2.8	5.4	4.5	11.8	6.7	12.5	–
All accidents	35.1	45.3	44.4	36.0	57.0	148.1	53.0	35.3	66.2	49.1	264.7	–
Malaria	–	0	–	8.2	–	–	–	3.6	–	–	–	–
Other arthropods	–	–	–	–	–	–	–	–	–	–	–	–
Diabetes	30.2	14.3	10.3	2.7	11.1	12.4	29.8	9.2	5.4	6.2	4.7	–
All nutritional	4.9	1.6	0.4	3.5	1.2	1.0	6.5	14.2	17.1	2.0	40.8	–

TABLE 1.7. Infectious disease notifications for 12 warm climate countries

	Fiji 78	Australia 78	Cuba 79	Thailand 79	Ecuador 79	Sri Lanka 78	Singapore 79	Guinea Bissau 79	Sudan 78	Saudi Arabia 78	Swaziland 78	Guatemala 79	Mean
Dysentery-enteritis	0.18	0.10	1.42	4.94	–	8.44	0.49	48.7	192.98	13.42	4.05	3.14	53.55
Malaria	–	0.02	0.03	1.68	1.05	5.16	0.09	291.01	153.12	18.81	1.49	10.43	43.44
All venereal infections	–	1.10	1.42	–	0.68	0.47	5.59	15.09	6.20	0.48	0.18	0.51	3.17
Influenza	0.003	–	–	0.72	1.92	3.79	–	7.41	1.88	–	0.19	9.18	3.14
All helminths	0.002	0.02	0	3.07	–	3.07	–	–	6.36	–	0.04	–	1.90
Schistosomiasis	–	–	0.0001	–	–	–	–	1.51	5.66	–	0.20	–	1.84
Measles	0.003	–	0.77	0.29	0.54	0.43	–	6.78	0.95	4.70	1.67	0.51	1.66
All tuberculosis	0.05	0.10	0.12	0.30	0.40	0.45	1.23	1.39	1.60	2.36	2.72	4.49	1.26
Whooping cough	0.003	–	0.02	0.11	0.25	0.05	–	2.49	6.21	1.11	0.03	0.22	1.15
Dengue-encephalitis	0.002	0.003	7.78	0.30	0.002	0.10	0.07	–	–	0.01	0.10	0.01	0.83
Infectious hepatitis	0.01	0.19	2.26	0.28	–	0.72	0.11	0.92	2.37	0.32	0.11	0.25	0.68
Filariasis	0.005	–	0.001	–	–	0.26	–	2.12	–	–	0.004	–	0.48
Cholera-Salmonella	0.005	0.15	0.02	0.24	0.41	0.67	0.09	0.01	0.32	0.08	0.15	0.19	0.19
Leprosy	–	0.004	0.04	0.01	–	0.05	0.03	0.70	0.30	0.01	0.17	–	0.14
Tetanus	0.002	0.001	–	0.04	–	0.14	–	0.52	0.11	0.01	0.05	–	0.12
Diphtheria	–	<.001	0.001	0.04	0.003	0.02	0	0.01	0.10	0.02	0.01	0.001	0.03
All poliomyelitis	–	<.001	<.001	0.02	0.001	0.01	<.001	0.02	0.03	0.03	0.01	0.01	0.01
All rickettsiae	–	<.001	–	0.001	0.008	–	–	–	–	–	0.002	0.002	0.002
Leishmaniasis	–	–	0	–	0	–	–	–	0.16	–	–	–	?
All trypanosomiasis	–	–	0	–	–	–	–	0.01	–	–	–	–	?

treated with caution. Reporting of new incidents of sickness are suspect even in technologically developed nations and at best these data reflect the frequency of such major infections as tuberculosis, leprosy, and syphilis where the major components of care are provided by the government as part of a regulatory programme.

Disease-related data serve a variety of purposes. For instance, the data allow comparisons between populations and signify what particular diseases are of special importance in the individual countries. The data also provide information on the progress of frontier-crossing epidemics or pandemics, and may indicate cyclic variations.

The mean figures for the 12 countries given in the final column of Table 1.7 confirm the diarrhoeal diseases as the principal problems even though the likelihood of under-reporting of such diseases is very high. This tendency for under-reporting is confirmed by comparing the figure of 19 298 diarrhoeal events per 100 000 in the Sudan with the 10 per 100 000 in Australia. The latter figure is most unlikely however favourable the environment. The high figures reported generally for sickness in the Sudan do not suggest it is an especially unhealthy country but merely that its reporting services are more active than in others with similar problems.

After diarrhoeal infections comes malaria with an overall mean of 4344 episodes per 100 000 inhabitants in a year, but the cases were concentrated in Guinea Bissau and the Sudan. This suggests that malaria is not perceived as so great a problem as the diarrhoeas in some tropical countries regardless of its actual frequency. No other typically tropical disease appears prominently in the listing of mean incidence in Table 1.7. This may in part reflect the chronic nature of diseases such as schistosomiasis and filariasis.

Even so, the mean figures hide variation between countries and diarrhoea is not always the principal reported sickness. Malaria was ahead in Saudi Arabia and Guinea Bissau. Influenza is a problem in Ecuador. In Singapore the perceived problem is venereal diseases and tuberculosis, whereas in Australia it is mainly venereal diseases. Thailand has remarkably low notifications in all infections although it adhered to the pattern of diarrhoeal infections first and malaria second.

The non-tropical diseases which appear most frequently are venereal diseases, influenza, measles, tuberculosis, whooping cough, and infectious hepatitis. Such disease distribution trends have been used to argue that the load of sickness in the developing countries is not so much a function of the climate *per se* as of poverty, ignorance, and low standards of hygiene. Care must be exercised in reaching such conclusions because the nature of the governmental regulatory processes plus the local interpretation of what is normal or acceptable may considerably distort what is actually reported.

However, it does seem that when the morbidity and mortality data are combined, the idea that the traditional tropical diseases are important as population health hazards is to some degree undermined. Sadly, although as noted by Robinson (1985) the survivors of the diseases of poverty face death by violence, while cancer and the degenerative diseases gain ground even as quite moderate economic development proceeds.

One inference to be drawn is that there is no typical 'developing' or 'warm climate' country. The vector-borne or parasitic diseases are present in different proportions as additions to the major burden of diseases which is common to all economically underdeveloped, educationally neglected populations regardless of climate.

Indeed, it can be argued that additional burdens are being added to those populations. One of the most significant changes this century has been the rapid growth of urbanization. While the world's population has doubled, the numbers living in cities have increased five-fold. The beginnings of economic development have additionally threatened previously stable rural populations and fostered migration and the formation of unstable urban communities with all their attendant high risks to life and well-being.

Disease and fertility

There are a number of ways in which disease status affects fertility, either by reducing fecundability, causing sterility, or by extending the birth interval (McFalls and McFalls 1984).

There is little dispute over the effect of severe food deprivation in reducing fertility. The Dutch famine of 1944–45 showed that births fell by 50 per cent some nine months after the famine (Stein and Susser 1978). A similar marked reduction in births was seen in Bangladesh after the floods and famines of 1974–75 (Mosley 1979).

It is unclear whether the decline in fertility was a result of irregular or even total cessation of the menstrual cycle which was brought about because of marked loss of weight, psychological stress, or reduced levels of intercourse. There is considerable evidence that a marked loss of weight is usually accompanied by an interruption of reproductive cycles (Frisch 1984). Frisch and Revelle (1971) suggested that the age of menarche was determined by attainment of a 'critical weight' but later studies indicated that an important factor was the proportion of body fat to lean body mass. The hypothesis that a minimum body weight and/or fat content is required for the onset of normal reproductive function also seems to be applicable to menstruating women who cease to experience cycles during episodes of weight loss (Lunn 1988). Women who experience marked reduction in body fat due to intensive exercise and training also cease reproductive cycles. However, women do resume menstruating when they are re-fed and approach Frisch's minimum weight and fat level.

Whilst there is general agreement that severe nutritional deprivation can affect maternal fecundity, the effect of moderate nutritional deprivation, which is the norm in many developing countries is less clear. In two Bangladeshi studies where maternal weight and weight for height were used as indicators of nutritional status there was no significant difference between the lightest and heaviest groups in length of postnatal amenorrhoea although the results were in the predicted direction of the lightest women having a longer period of amenorrhoea. Studies of the effects of maternal depletion on fertility are complicated because of the interaction between nutritional status, lactation, and breast-feeding patterns, which all play a role in the return of ovulation.

Maternal nutritional status has also been shown to affect birth weight and there is a wealth of data which demonstrates that low birth weight is an important predictor of neonatal and infant mortality, child growth, and development, and the future reproductive performance of newborn girls (Martorell and Gonzalez-Cossio 1987).

Infertility is a major problem in sub-Saharan Africa (Belsey 1976) and in 1976 the WHO reported on a programme in 17 countries which had used a standardized approach to the investigation and diagnosis of infertile couples. A total of 8343 couples participated but there were no controls. The major categories of diseases affecting fertility for females are shown in Table 1.8.

Sexually transmitted diseases are a main cause of tubal disease. In adults in Africa, gonorrhoea is probably the most common communicable disease after colds, malaria, and gastroenteritis. In females, the most important complication is salpingitis. It commonly arises during the first or second menstrual cycle after the infection and it is estimated that 20 per cent or more of women with cervical gonorrhoea in African countries develop salpingitis. The infection leads to tubal occlusion, ectopic pregnancy, persistent lower abdominal pain, and infertility. In males, it leads to epididymitis and among men over 30 years of age, about 45 per cent of those with bilateral epididymitis were childless.

Other infectious diseases can also affect fertility. For instance, genital tuberculosis in females can cause sterility by the formation of tubercules in

TABLE 1.8. Major categories of diseases affecting fertility (%)

No demonstrable cause	27
Tubal disease	31
Ovulatory disorders	24
Other	18
All categories	100

TABLE 1.9. Mortality rates for selected causes in parous and nulliparous women aged 45 to 74, England and Wales 1959 −60 (1)

Cause of death	ICD (2)	No. of deaths	Relative mortality rates parous/nulliparous (%)
All causes		101 262	120
Cancer of cervix	171	2 150	139
Heart disease	420−422	17 543	138
Hypertensive disease	440−447	3 349	137
Cerebrovascular disease	330−334	16 324	128
Cancer of breast	170	7 131	92
Cancer of endometrium	172	852	74
Cancer of ovary	175	2 461	73

(1) Figures abstracted from Tables 1 and 2 in Beral, V. long term effects of childbearing in health. *Journal of Epidemiology & Community Health*, 39, 343−6.
(2) WHO (1948). *Manual of the international statistical classification of diseases, injuries, and causes of death.* World Health Organization, Geneva.

the Fallopian tubes. Malaria can induce pregnancy wastage and schisto-somiasis may affect fertility by producing lesions in the genital tract.

Fertility also plays a major aetiological role in a number of diseases (Pike 1988). Comparisons of parous and nulliparous women in England and Wales showed that during the period 1959−60 parous women had a 20 per cent increased mortality rate (Table 1.9). This increase is largely due to four diseases, cancer of the cervix, coronary heart disease, hypertension, and cerebrovascular disease. Three diseases are consistently reduced in parous women, cancers of the breast, endometrium, and ovary. Six of the seven diseases have been linked to the use of oral contraceptives.

Increased parity decreases to a very significant extent the risk of cancers of the endometrium, ovary, and breast. Pike has estimated that 5 years of oral contraceptive use will reduce the lifetime risk of endometrial and ovarian cancers by about 40 to 50 per cent, and 10 years use will reduce that risk by 65 to 75 per cent; a significant reduction in morbidity and mortality. Oral contraception has not been found to provide any protection against breast cancer and there is some weak evidence that the risk may be increased by early oral contraceptive use.

Disease, productivity, and work capacity

Many studies on productivity and/or work capacity have focused on infectious diseases in developing countries. The effect of disease should be related to the demand/supply of labour. In many parts of Africa in the dry season there is an excess of labour and the incapacity of one family member will have little economic effect. However, in the planting and harvesting seasons the demands on labour are high and therefore illness may have a serious economic effect.

Some chronic diseases, such as onchocerciasis (river blindness) and deformities from poliomyelitis, make the individual less productive but it is not easy to measure the effects of ill health due to tuberculosis, leprosy, and chronic anaemia (Parry 1984).

However, an Indian tea plantation study (Bradley et al. 1988) showed a clear relationship between anaemia and productivity. A total of 240 women were divided into four groups A1, A2, B1, and B2. A1 and A2 received iron tablets; B1 and B2 a placebo. In addition, A2 and B2 received a single course of worm treatment. The results over a 6 month period showed a rise in haemoglobin level of + 1.7 g/100 ml and in weight + 2.6 kg among those receiving iron and a non-significant fall in these measures among the placebo group. The group receiving only iron supplements showed a rise of 2 days a month in the time worked and an increased yield of 92 kg/month, a 22 per cent rise in productivity, whereas over the same period the two groups not receiving iron showed a 45.5 kg/month fall in yield, probably because of the difference in season as compared with the baseline observations. Those who received worm treatment and iron showed a small rise in yield (which is not fully explained!), although their change in yield was much more favourable than for the two groups not receiving iron. Those female workers who had received iron strongly emphasized that they had a sense of well-being that was not reported by those who received a placebo.

Bradley et al. (1988) also examined the effects of pregnancy on plucking performance. They showed that productivity was not affected until the women went on maternity leave 6 to 8 weeks before delivery. Women took on average 100 days leave as against an authorized 82 days, whereas primiparous women took more leave than women in subsequent parities. They attained their average plucking performances in the 5th month after returning to plucking. Then performance went down from the 6th to 8th month and picked up in the 10th and 11th month. They maintained their averages once again in the 12th month.

Another water-borne disease which has been studied for its effect on agricultural productivity is the Guinea worm (Dracunculus medinensis). The adult female worm migrates to subcutaneous tissues particularly the feet and legs. A toxin is then released leading to the formation of painful ulcers. Nwosu et al. (1982) in a study of four heavily infected villages in the Western State of Nigeria found that nearly half the men of working age were incapacitated for at least 10 weeks. In another study in northern Uganda, Guinea worm disabled 75 per cent of men and women in a village during the cotton-picking season. In southern India the infection disabled men for about a month. Belcher et al. (1975) found a high worm burden in northern Ghana farmers. Their infection allowed them to do domestic duties but they were often unable to farm for five or more weeks.

Schistosomiasis is another disease in which the relationship between physiological fitness and infection has been studied. In savannah countries where the economy depends on manual labour and the prevalence of schistosomiasis is high, the loss of productivity has been estimated to be up to 40 per cent. Until recently objective measurements have not been made and so it is questionable whether infection results in a deterioration in the capacity to work.

Parker (1989) studied the effects of infection with *Schistosoma mansoni* on female activity patterns in the cotton fields and the domestic sphere. Her elegant study, in the Gezira province of the Sudan, was based on matched pair analysis of infected and uninfected women. The matching took into account a wide range of social and economic factors. Although the samples were small (22 women, 11 pairs in the cotton fields and 24 women, 12 pairs in the domestic sphere) the results were illuminating.

Infected women, those with a mean egg load of 1462/g, attempted to pick as much cotton as possible in the shortest possible time in the morning work session. In the afternoon this pattern was partially repeated but a significant number of infected women felt too weak to work. In the domestic sphere, infected women (mean load of 574 eggs/g) did not spend significantly less time engaged in domestic activities than uninfected women. No significant relationship was found between intensity of infection and the nutritional status of women and their infants.

A number of studies have attempted to examine the effect of *S. mansoni* on physiological tests of work performance and a smaller number have been conducted with *S. haematobium*. The results are conflicting. For instance, Omer and Ahmed (1974) studied physiological performance using a Harvard Step Test and found significant differences between infected (with *S. mansoni*) and uninfected male nurses as well as in pre- and post-treatment performance. However the infected nurses had no hepatic or splenic enlargement. Collins *et al.* (1976) examined the physiological performance of 194 Sudanese cane cutters in the age groups 16–24 and 25–45. Each group was divided into three clinical groups comprising those with no infection, presence of infection, and infection with hepato-splenomegaly. Unfortunately, egg loads were not recorded. Physical performance was measured using a stationary bicycle ergometer. No significant differences in physiological performance were found between those not infected and those infected to different degrees with *S. mansoni*.

These two studies are clearly not comparable. The physiological tests were different and it is to be expected that the level of physical fitness and muscular power of cane cutters was higher than nurses. Both studies were biased in favour of fitter individuals since the studies were obtained from volunteers which would automatically exclude the sick, less fit individuals.

Awad el Karim *et al.* (1980) examined the effects of *S. mansoni* on exercise performance of 203 male villagers and canal cleaners living in four villages in the Gezira, Sudan. They found that egg loads of *S. mansoni* up to 1500/g faeces only had a small effect on exercise performance but with egg loads above 2000/g, maximum aerobic power could be reduced as much as 20 per cent, i.e. physical capacity was impaired only in those with high intensity of infection as assessed by egg load.

However, the experimental group (canal cleaners) and controls (villagers) differed markedly in their daily activity patterns, income and haemoglobin levels. The canal cleaners had lower haemoglobin and income levels (half that of villagers) and higher activity patterns. Thus, it is unclear to what extent the differences in physical performance reflected depleted haemoglobin, nutritional, and/or infection effects.

There is some evidence which suggests that *S. haematobium* infection decreases the physical fitness of children. Stephenson *et al.* (1985, 1986) matched infected and uninfected Kenyan children for potential confounding effects. They showed that infected children were physically less fit and had higher heart rates after completion of the Harvard Step Test than uninfected children.

The exact mechanisms by which *S. haematobium* and *S. mansoni* affect physical fitness are unclear. Urinary schistosomiasis (*S. haematobium*) could produce iron deficiency, iron deficiency anaemia, and protein-energy malnutrition (Stephenson 1987). *Schistosoma mansoni* causes blood loss in the stool, but the magnitude of blood loss over time and its significance for nutritional status is unclear. In seven chronically infected Egyptian patients the blood loss averaged 12.5 ml/day which is equivalent to 3.3 mg of iron per day. Three of the seven patients were severely anaemic.

In spite of the differences in research design and methodologies the results suggest the high intensity of infections of *S. mansoni* and possibly *S. haematobium* affect physiological performance of active male workers.

Parker's results suggest an economic component to *S. mansoni* infection. The early studies reviewed by Prescott (1979) indicate considerable economic benefit would accrue from reducing the prevalence of *S. mansoni* and *S. haematobium*. For instance, Khalil (1949) calculated that £80 000 per annum was lost because of schistosomiasis in Egypt while Wright (1951) came up with the figure of a fall in productivity of 33 per cent due to schistosomiasis. Although these figures have been criticized for a number of reasons (see Weisbrod *et al.* 1973 and Gwatkin 1984) the extent to which schistosomiasis impairs labour productivity is still unresolved.

Attempts have also been made to examine the relationship between *S. mansoni* and mental development, school performance, and school absenteeism. This issue has been studied, often with inadequate study designs, and again the results are conflicting (Stephenson 1987).

Disease, nutrition, and growth

A loss of appetite and decreased tolerance for food are among the earliest and most constant results of infectious disease. Another practice especially common in the feeding of young children in less developed countries is the dietary change to a more liquid intake higher in carbohydrate at the expense of foods that are good sources of protein and other essential nutrients. Often when a child has diarrhoeal disease or other acute infection, milk and solid foods are eliminated in favour of starchy gruels or cooking water from cereal grains or plant leaves. Fever increases both basal metabolism and the loss of nutrients in sweat. Acute infection results in mobilization of amino acids from skeletal muscle and other tissues for glucogenesis in the liver. This process depletes body protein because the amino acids are deaminated to provide the carbon atoms in glucose, and the nitrogen is excreted in the urine largely as urea.

Nitrogen loss has been recognized during severe infectious diseases of bacterial origin, such as typhoid fever and tuberculosis, and can be equivalent during the acute phase to the nitrogen of 2 to 3 kg of muscle. Protein energy-deficient diseases, kwashiorkor and marasmus, are commonly precipitated by acute diarrhoeal disease, measles, whooping cough, and other childhood infections in persons whose nutritional status is precarious. Similarly infections, especially chronic infections, are responsible for much of the protein energy deficiency seen in adults.

Blood levels of vitamin A are reduced in acute infections and xerophthalmia and keratomalacia frequently follow in children whose diets are deficient in the vitamin. Pneumonia and malaria have been recognized as significant factors in the occurrence of beriberi (as a result of thiamin deficiency) in civilian populations, and dysentery was a common precipitant of the disease among prisoners of war during World War II held by the Japanese.

Infectious or serum heptatitis cause raised levels of serum triglycerides and cholesterol and absorption of fat is decreased in infections that provoke diarrhoea. Chronic infections of any type are likely to produce so-called anaemia of infection by shortening the life span of erythrocytes and interfering with red cell production in bone marrow. Two of the most common infectious diseases, weanling diarrhoea and measles, have often been used to illustrate the way in which infection and nutritional deficiency combine synergistically (see Fig. 1.6). Measles is more severe in younger children probably in part because they are more liable to be undernourished. Even if the children are well nourished at the start of the illness their nutritional

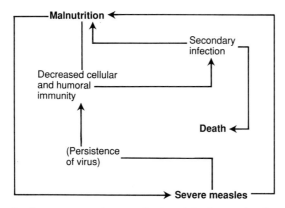

Fig. 1.6 Measles demonstrates the complexity of the interaction between nutrition and infection.

status is much more labile and within a week may be deteriorating so that they enter the cycle shown in Fig. 1.6.

Intestinal helminths have been implicated as a factor contributing to human malnutrition. Helminths reside in the alimentary tract where digestion and absorption of nutrients takes place. Damage to the intestinal mucosa, interference with nutrient absorption, and greater loss of nutrient through the damaged intestine may all contribute to reduced nutritional well-being of the human host. Reductions in food intake by the host due to the presence of the helminth infection could also contribute to poor nutritional status.

Ascaris infection has long been implicated in the aetiology of malnutrition including kwashiorkor. Intestinal tracts from *Ascaris*-infected children and pigs show certain abnormalities. Tripathy *et al.* (1972) found that jejunal biopsies from infected children showed broad and shortened villi, elongation of crypts, and a decrease in the villus–crypt ratio.

Absorption of nutrients may be impaired by *Ascaris* infection. Some clinical trial results showed reduced nitrogen, fat and xylose absorption in subjects infected with *Ascaris*, primarily when the studies were carried out before and after de-worming. However, the effects were not large and the nutritional significance would probably only be important in individuals consuming marginal diets.

Vitamin A utilization has been found to be reduced in children infected with *Ascaris*. Clinical studies by Mahalanabis *et al.* (1979) and by Sivakumar and Reddy (1975) have shown that absorption of an oral dose of vitamin A was significantly reduced in children infected with *Ascaris* compared with uninfected control children. There are also reports of a greater incidence of xerophthalmia and reductions in serum vitamin A in children infected with *Ascaris*. A large-scale field study in Panama showed that infected children had lower serum vitamin A and carotene values compared to uninfected children.

Many of the effects reported on digestion and absorption of nutrients have been shown to be related to the intensity of infection. The number of worms recovered following administration of an anthelmintic drug are highly variable between individuals. Generally, the results conform to a negative binomial distribution (Anderson 1985) where populations have relatively few individuals infected with large numbers of worms and as many as 65 per cent of the parasites can be harboured by less than 15 per cent of the individuals. The more heavily infected individuals may be expected to show larger reduction in nitrogen and fat absorption, vitamin A utilization, and lactose digestion and this has generally been observed (Forsum *et al.* 1981; Carrera *et al.* 1984; Taren *et al.* 1987).

Several authors have reported a significant association in field situations between *Ascaris* infection and growth. The growth studies have recently been reviewed by Brown and Gilman (1986) and are shown in Table 1.10. Three of the studies show improved nutritional status as measured by weight gain or skinfold thickness for children uninfected with *Ascaris*, or children treated with an anthelmintic drug compared to infected children.

Foo (1986) found a negative relationship between the intensity of *Ascaris* infection and child height, although not with weight and skinfold thickness. In an intervention study Foo went on to show that children freed of *Ascaris* infection by anthelmintic treatment gained more weight and had greater skinfold thickness compared with *Ascaris*-infected children over a one year study. The difference observed, 327 g between *Ascaris*-infected and successfully treated children was about 16 per cent of the average one year increment in worm-free children.

Studies examining the influence of *Ascaris* infection on child growth have not produced unequivocal results. The conflicting results could be due to a failure to address the issue of intensity of infection relative to effects on child growth. Cerf *et al.* (1981) in a Bali study found the effect on child growth primarily in those with heavy infections.

In addition, different anthelmintic drugs have been used, and some children have polyparasitism of *Ascaris*, *Trichuris*, and hookworm. Differences also exist in sampling methodologies and at what age children are studied. These differences as well as failure to control for socio-economic variables and other aspects of health status probably account for most of the apparent conflicting results.

In one Bangladesh study children were regularly de-wormed using a broad spectrum anthelmintic which is active against *Ascaris*, *Trichuris*, and hookworm. Their anthropometric measurements were compared with a

TABLE 1.10. Comparison of five field studies of increments in nutritional status following treatment of ascariasis in children

Country	Type of therapy	Control group	Length of follow-up of nutritional status	Nutritional status outcome
India	Tetramisole, one daily dose 2 days, every 4 months	Assigned to therapy or placebo by village	1 yr	Greater proportion of treated children showed improved nutritional status
Tanzania	Levamisole, 3 monthly	Matched *Ascaris*-infected and non-infected control (placebo)	1yr	Greater weight gain in levamisole-treated, ascaris-infected group than in matched controls
Kenya	Levamisole, single dose	Uninfected children	14 weeks before therapy, 14 weeks after therapy	Poorer increments of triceps fatfold in infected group before therapy and better increment of triceps fatfold and weight in infected group after therapy
Bangladesh	Piperazine, two doses within 2 week period	Random assignment to placebo	11 months	No effect of therapy. (Although incomplete cure rates and rapid reinfection were noted, no nutritional effect was observed even when only cured patients were considered)
Brazil	Mebendazole, two doses × 3 days	Random assignment to placebo	10 months	No effect of therapy on nutritional status even when only cured patients were considered

Source: From Brown and Gilman (1986) with permission.

control group given a placebo. Over a one year study no significant differences in height for age, weight for age, and weight for height were detected. However, the incidence of *Giardia lamblia* rose from 4 per cent to 30 per cent in the de-wormed group whereas no increase in *Giardia* incidence was observed in the control group. *Giardia* is known to cause diarrhoea and one possibility why children did not show weight and height gains following de-worming was increased diarrhoeal episodes and malabsorption caused by *Giardia*.

About 1 billion people world-wide are thought to be infected with hookworm, and *Ancylostoma duodenale* and *Necator americanus* are the two primary species infecting human populations. Hookworms live attached to the intestinal mucosa and feed on mucosal tissue, blood, and other tissue fluids. Blood loss occurs during feeding and this loss, facilitated by the feeding activity of hookworms is considered to be responsible for iron deficiency anaemia, the principal nutritional consequence of hookworm infection.

The association of hookworm infection and anaemia has been observed by many investigators (e.g. Layrisse and Roche 1964; Roche and Layrisse 1966). The consequences of iron deficiency anaemia may be reduced work capacity, especially for those engaged in heavy work. Other nutritional consequences of hookworm are less clear and the relationship between hookworm infection and child growth rates are equivocal. The majority of the work shows that infection with hookworm is not associated with malabsorption syndrome.

Human infections with *Trichuris trichuria* (whipworm) are nearly as common as infection with *Ascaris lumbricoides* and hookworm. However, the *Trichuris* parasite resides in the large intestine and thus its nutritional effect might be expected to be less than that due to *Ascaris* or hookworm. Children suffering from heavy *Trichuris* infection show signs of chronic diarrhoea, anaemia, and growth retardation (Bundy 1986; Holland 1987) and several studies of children with heavy infections with *Trichuris* have reported that such children are underweight and often shown signs of severe malnutrition including marasmus and kwashiorkor.

Schistosomiasis is also thought to affect adversely human nutritional status. *Schistosoma haematobium* causes haematuria and proteinuria, and daily blood loss varies between 0.5 ml and 125 ml per day. Hospitalized patients with severe or persistent haematuria were losing a mean of 22 ml blood per day. Blood losses of this magnitude will, if not compensated for by increased dietary iron intake, lead to anaemia.

Stephenson *et al.* (1985, 1986) conducted a study of urinary iron losses due to *S. haematobium* in non-hospitalized children in Kenya. The children were divided into three groups, uninfected, low to medium egg counts, and high egg counts. The children were examined before and after treatment.

Before treatment the control group was losing 149 μg of iron per 24 hours, while in the low to medium group the loss was 278 μg, and in the high group 652 μg. There was a simple linear relationship between log of iron loss and log of egg count ($r = 0.56$, $P < 0.0003$). After treatment iron losses were similar in all three groups.

Six studies in Africa suggest that urinary schistosomiasis may inhibit child growth but after treatment growth rates improve. However, four other cross-sectional studies failed to find any significant relationship between S. haematobium infection and child's anthropometric measurements. In three of these studies intensity of infection was not considered. These studies highlight the need to measure the intensity of infection, with longitudinal rather than cross-sectional studies, preferably controlling for age, sex, diet, and socio-economic status.

Stephenson conducted a study which examined the relationships between hookworm, S. haematobium, and malarial infections, and child growth of Kenyan children over a 6 month period. Infected children with egg counts less than 500 eggs/10 ml were allocated to either treated ($n = 201$) or placebo ($n = 198$) groups. The treated group gained significantly more weight (0.8 kg), arm circumference (0.4 cm), percentage arm circumference for age (1.7 per cent), and in triceps and subscapular thicknesses. The placebo group showed significant decreases in percentage weight for age, percentage arm circumference for age, and both skinfold thicknesses for age over the 6 month period.

The exact mechanism by which treatment for S. haematobium infection improves growth rates in children is still unknown. Stephenson has suggested that proteinuria is not the major mechanism but that food intake decreases during infection due to anorexia and subclinical morbidity.

The relationship between S. mansoni and malnutrition has also received research attention and nine population based cross-sectional studies have enquired whether infection is related to anaemia, growth, physical fitness, and other indicators of nutritional status, such as serum albumin. The results are equivocal, with some studies showing significant effects of infection and others not. However, well-controlled animal studies have shown that S. mansoni infection causes anaemia, anorexia, and weight loss. The same limitations apply to S. mansoni as were detailed for S. haematobium.

In summary, the difficulty in determining the relationship between many helminths and malnutrition is further complicated by the association with other parasitic infections, malnutrition, poverty, and poor hygiene conditions.

The effect of nutrition on non-infectious disease is of increasing interest particularly the relationship between vitamin A and lung cancer, dietary fat and breast cancer, and diet and coronary heart disease. Vitamin A plays a central physiological role in the regulation of cell differentiation. Because loss of differentiation is a basic feature of malignancy, vitamin A may be related to cancer incidence. A number of animal studies have shown that preformed vitamin A inhibits the occurrence of induced tumours and even reversed metaplastic changes (Hill and Grubbs 1982; Sporn and Roberts 1983).

Interest in vitamin A as a potential inhibitor of human cancer increased following Bjelke's (1975) study of 8278 Norwegian men. After adjusting for the effects of cigarette smoking, Bjelke found that the rate of lung cancer among men whose calculated intake of vitamin A was above average was only one-third that of men with intakes below average (Table 1.11).

The interpretation of Bjelke's finding was complicated by the diversity of vitamin A sources which can be preformed as carotenoid. A second study on this Norwegian sample suggested that the protective effect against lung cancer was primarily attributable to carrots and other vegetables with some additional contribution from milk. This suggests a beneficial effect of carotenoid sources of vitamin A rather than for preformed vitamin A. This finding was supported by a 19-year follow-up of 2107 men enrolled in the Western Electric Study (Shekelle *et al.* 1981) which showed that preformed vitamin A was not related to the incidence of lung cancer, whereas intake of vitamin A from carotenoids was strongly associated with lower risk of the disease. Two other studies, one in Japan (Hirayama 1979) and the other in the United States (Long-de and Hammond 1985) are generally supportive of the earlier studies. Seven case-control studies have been undertaken and they provide support for the protective relationship between carotenoid sources of vitamin A and the occurrence of lung cancer.

Studies have also been undertaken to examine the relationship between blood vitamin A levels and the risk of lung cancer. However, the body of evidence relating serum retinol levels to risk of lung cancer as well as overall cancer incidence is overwhelmingly null. Attempts have recently been made to study the relationship between blood levels of beta-carotene and risk of lung cancer. (This relationship has been difficult to study hitherto for two reasons: assays to measure beta-carotene in large numbers of blood specimens have only recently become available; and beta-carotene is far less stable in frozen sera than retinol and ultra-low temperatures (-80 °C) are needed to avoid degradation). The results generally suggest that higher serum beta-carotene levels are associated with lower risk of lung cancer.

The interpretation of data on blood levels of beta-carotene and lung cancer has been complicated by recent suggestions that cigarette smoking may have a metabolic effect that reduces blood levels of beta-carotene. Although blood levels of beta-carotene are generally lower in smokers than non-smokers this finding could simply reflect the reduced dietary intake of fruits and vegetables by smokers but this suggestion has been refuted by recent studies (Russel-Briefel *et al.* 1985; Stryker *et al.* 1988).

TABLE 1.11. Age-adjusted rates of lung cancer among men with high vs. low vitamin A index

	Rate of lung cancer (1)		Relative risk
	Vitamin A index ≤ 5	≥ 5	Vitamen A index ≥ 5 vs. < 5
Cigarette smoking status			
Ever smoked	10.6	4.2	0.40 (2)
Current smoker, >20 cigarettes/day	21.0	7.4	0.35
Current smoker, 1–19 cigarettes/day	12.8	5.7	0.44
Ex-smoker	6.1	1.5	0.25
Never smoked	1.1	1.2	1.01
Total, smoking adjusted	7.3	2.8	0.38 (3)

(1) Age-adjusted 5-year cumulative incidence/1000 men.
(2) $p < 0.05$.
(3) $p < 0.01$.
Data are based on 53 cases occurring during a 5-year follow-up of 8278 Norwegian men. From Bjelke (1975).

If these results are confirmed then this implies that smoking has a direct metabolic effect on blood beta-carotene and renders those studies of blood beta-carotene and lung cancer risk, viewed on their own, almost uninterpretable (Willett 1990). However, other evidence suggests that the higher risk of lung cancer among individuals with low serum beta-carotene is not merely due to a metabolic effect of smoking on the blood levels, although it is possible that the magnitude of the relationship with serum levels may be exaggerated. The inverse relationship between intake of vegetables and fruits and risk of lung cancer which has been found in both case-control and cohort studies represents one of the best-established associations. What is less clear is whether this is due to a causal effect of beta-carotene.

The relationship between diet and coronary heart disease has already been discussed in some detail (see Section 2). Recently, the National Research Council of the United States recommended that the fat content of the US diet be reduced from an average of 40 per cent to 30 per cent of calories based on an anticipated reduction in breast cancer rates.

Nearly 40 years ago, Tannenbaum and Silverstone (1953) showed that the amount of dietary fat could markedly influence the occurrence of mammary tumours in rodents. It was the publication by Armstrong and Doll (1975) on the striking correlations among countries between the national per capita fat consumption and both incidence and mortality rates of breast cancer that finally attracted widespread attention. The international differences in breast cancer are particularly great for post-menopausal women and this information has been used to support the dietary hypothesis.

The international correlation is mainly with animal ($r = 0.83$) not vegetable fat ($r = 0.18$). The association is, of course, confounded by lean body mass, obesity, sedentary life style, and by any of the many other correlates of economic development. The high-incidence countries tend to be Westernized and the low-incidence countries non-industrialized.

Migrant studies have demonstrated that the large differences in breast cancer between countries are not primarily attributable to genetic factors. For instance, offspring of immigrants from Japan to the United States, but not the immigrants themselves, have breast cancer rates that are similar to those in the general American population (Buell 1973). The delayed effect among Japanese-Americans may be due to a slower acculturation process in Japanese than other nationalities, because Polish women who migrate to the United States and Italian women migrating to Australia attain breast cancer rates that are similar to the higher rates among women born in those countries.

Dramatic changes in breast cancer incidence have occurred in Iceland during this century (Bjarnason et al. 1974) which also provides strong support for non-genetic factors in the aetiology of the disease.

Some studies have been conducted on people consuming unique diets for long periods. The Seventh Day Adventists consume relatively small amounts of meat and other animal products (Phillips et al. 1980) and it was originally reported that their breast cancer mortality rates were lower compared to the general United States population. However, the mortality rate differences could be confounded by socio-economic status. In a British study Kinlen (1982) compared rates of breast cancer among orders of nuns who were either vegetarians or ate only small amounts of meat with rates among single British women. This study, which controlled for parity, revealed no significant difference in the rates of breast cancer.

Case-control and cohort studies have generally failed to confirm the hypothesis that a diet high in total or saturated fat composition increases the incidence of human breast cancer. Thus, the large differences in breast cancer rates among countries remains unexplained. Many alternative hypotheses exist including differential intakes of selenium and other minerals, marine oils, alcohol, and specific vegetables. In addition, late menopause, early menarche, and late full-term pregnancy have also been suggested as being able to explain part of the international variation in rates.

Disease and genetic variation

Haldane is generally regarded as the person who first discussed the evolutionary relationship between natural selection and disease. Haldane (1956) wrote: 'Infectious diseases have doubtless spread a great many previously rare genes through human populations. The fact that they were rare means that they mostly lowered the fitness of our ancestors. They probably lower our own, except in environments where the organism against which they confer resistance is fairly common. This is certainly true of the gene for sickle cell anaemia.'

Haldane was referring to the sickle-cell gene (HbS) which is found in relatively high frequencies where falciparum malaria is also present. In the homozygous form (SS) individuals suffer from sickle-cell anaemia which is frequently lethal. It is thus suspected that the maintenance of the HbS allele is due to the selective advantage of HbAS heterozygotes (sickle-cell trait) compared to homozygous (AA) normal individuals, i.e. there exists a balanced polymorphism.

Ten studies have been undertaken in Africa which have examined incidence of malaria in young children with and without sickle-cell trait (Table 1.12). Eight of the ten studies show significant differences with higher incidence of parasites in the non-sickle-cell trait children. When the incidence of heavy *Plasmodium falciparum* ($>1000/\mu l$) is compared in sickling and non-sickling subjects the differences are even more striking (Table 1.13) with significantly higher incidence of heavy infections in the non-sickle-cell subjects.

Deaths from malaria also support the notion that sickle-cell carriers are more likely to survive in malarious environments than non-sickling children (Table 1.14). However, there has been some discussion as to whether resistance against malaria is by differential viability or differential mortality. Because it is difficult to demonstrate any effect of the sickle-cell trait on malaria in adults Allison (1954) concluded that the main selective effect was exerted by differential mortality in young children before appreciable immunity against malaria developed. Direct evidence in support comes from studies carried out in West Africa, Kenya, and the Congo.

Livingstone (1957) suggested that differential fertility among women occurred as a result of placental malarial infection. Rucknagel and Neel (1961) worked out that in a population with 35 per cent sickle-cell trait carriers the equilibrium would be attained when the fertility of sickle-cell trait carriers (both sexes) was 1.97 times that of normal subjects and fertilities are additive in the sexes. Allison (1964) summarized the African data and found no evidence that the fecundity of sickle-cell trait carriers was higher. However, among the Black Caribs of Honduras the fertility ratio of sickle-cell trait to normal mothers was 1.45 to 1 which is sufficient to

TABLE 1.12. *Plasmodium falciparum* parasite rates in African children

Subjects and age in years	Sickle-cell trait		Non-sickle-cell trait		Relative incidence (1)	Weight	Woolf χ^2	Probability
	P. falciparum	Total	*P. falciparum*	Total				
Uganda, <6	12	43	113	247	2.18	7.58	4.60	$0.05 > p > 0.02$
Kenya, <6	131	241	154	241	1.49	28.81	4.53	$0.05 > p > 0.02$
Uganda, <10	73	191	494	1,009	1.55	38.26	7.36	$0.01 > p > 0.001$
S. Ghana	42	173	270	842	1.47	27.10	4.05	$0.05 > p > 0.02$
N. Ghana, <5	11	15	165	177	5.00	2.32	6.01	$0.02 > p > 0.01$
Nigeria, <5	162	213	680	890	1.02	31.24	0.01	$p > 0.99$
N. Ghana, <6	13	19	109	127	2.79	3.24	3.42	$0.10 > p > 0.50$
Nigeria, <6	51	91	245	342	1.98	16.95	7.93	$0.01 > p > 0.001$
Tanganyika, <5	77	136	272	407	1.54	24.38	4.60	$0.05 > p > 0.02$
S. Ghana	34	123	176	593	1.10	20.52	0.20	$p > 0.50$

(1) Incidence of *P. falciparum* infections in non-sickle-cell trait groups relative to unity in corresponding sickle-cell trait groups.
Weighted mean relative incidence = 1.46.
Difference from unity, $\chi^2 = 29.2$ (1 df), $p < 0.001$.
Heterogeneity between groups, $\chi^2 = 13.5$ (9 df), $0.20 > p > 0.10$.

TABLE 1.13. Incidence of heavy *P. falciparum* infections in African children

Country	Classification of infection	Sickle-cell		Non-sickle-cell		Relative incidence (1)	Weight	Woolf χ^2	Probability
		Heavy infections	Total	Heavy infections	Total				
Uganda	Group 2 or 3	4	43	70	247	3.86	3.38	6.16	$0.02 > p > 0.01$
Kenya	Heavy	21	241	38	241	1.96	11.99	5.43	$0.02 > p > 0.01$
Uganda	$>1000/\mu l$	35	191	374	1,009	2.63	25.49	23.74	$p < 0.001$
S. Ghana	$>1000/\mu l$	3	173	57	842	4.11	2.79	5.59	$0.02 > p > 0.01$
N. Ghana	$>1000/\mu l$	5	15	75	177	1.47	3.07	0.46	$p > 0.50$
Nigeria	$>1000/\mu l$	25	91	147	342	1.99	14.91	7.06	$0.01 > p > 0.001$
Tanganyika	$>1000/\mu l$	36	136	152	407	1.66	20.71	5.27	$0.05 > p > 0.02$
S. Ghana	$>5630/\mu l$	3	123	42	593	3.05	2.72	3.38	$0.10 > p > 0.05$

(1) Incidence of heavy *P. falciparum* infections in non-sickle-cell trait groups relative to unity in corresponding sickle-cell trait groups. Weighted mean relative incidence = 2.17. Difference from unity $\chi^2 = 51.379$ (1 df), $p < 0.001$. Heterogeneity between groups $\chi^2 = 5.719$ (7 df), $0.7 > p > 0.5$.

TABLE 1.14. Deaths from malaria in relation to the sickle-cell trait in African children

Reference	Subjects	No. of deaths	No. with sickle-cell trait	Incidence of sickle-cell trait in population	Probability
Raper (1956)	Uganda (Kampala)	16	0	0.20	0.028148
J. and C. Lambotte-Legrand (1958)	Congo (Leopoldville)	23	0	0.235	0.0021095
Vandepitte (1959)	Congo (Luluaborg)	23	1	0.25	0.115938
Edington and Watson-Williams (1964)	Ghana (Accra)	13	0	0.08	0.33826
Edington and Watson-Williams (1964)	Nigeria (Ibadan)	29	0	0.24	0.00034953

$\chi^2 = 46.4$ (10 df), $p < 0.001$.

maintain the sickle-cell gene even in the absence of differential mortality due to malaria.

The exact cellular mechanism whereby HbS confers protection is still unclear. Either the merozoites fail to invade HbAS cells or the parasites do not grow and divide in these cells, or the parasitized HbAS red blood cells are more rapidly removed by the host's defence system. It has also been suggested that P. falciparum causes HbAS red cells to sickle in vivo and that these cells are preferentially phagocytosed.

Experiments in vitro were carried out at normal and reduced oxygen tension. They showed that the invasion rate of HbAS, AA, and SS red cells by merozoites was about the same at normal oxygen tensions, but in reduced tensions the invasion rate was significantly depressed in HbSS cells and to a lesser extent in HbAS cells.

Other haemoglobin forms have also been thought to associate with malaria. Haemoglobin C is found in a small area of West Africa. Homozygous (HbCC) individuals have only a mild haemolytic anaemia but the growth of P. falciparum in culture was adversely affected in HbCC erythrocytes. Haemoglobin F or fetal haemoglobin (HbF) may also confer some selective protection particularly on children with beta-thalassaemia during infancy and early childhood. This is because in these children HbF production persists longer than usual. Children who are homozygous for beta-thalassaemia die early. However, the high frequency of this genetic condition in some formerly malarious areas of Greece and Sardinia suggests that beta-thalassaemia confers some resistance to malaria (Bruce-Chwatt 1980).

It has also been shown that malaria parasites in glucose 6-phosphate dehydrogenase deficient (G6PD) cells are more easily damaged by the effects of oxidants than they are in non-G6PD deficient cells. The enzyme deficiency is X-linked but appears to be harmless unless the red cells are challenged in some way, usually by exposure to various drugs, for instance primaquine.

In order to invade a red blood cell the merozoite must attach itself to the cell surface membrane through receptors on the cell surface. These receptors are glycoproteins and are species specific. Miller (1977) showed that individuals lacking the Duffy blood group antigens (Fy[a] and Fy[b]) are refractory to invasion by P. knowlesi. These antigens are absent in much of Africa where P. vivax infections are virtually unknown. It was thus postulated (Livingstone 1984) and later proved that this antigen was related to a receptor for P. vivax. However, Duffy negative red cells are susceptible to P. falciparum.

Bruce-Chwatt (1980) summarized the three possible mechanisms by which a genetic characteristic of the host red blood cell can increase natural resistance to malaria as: (1) surface receptors of the red blood cell can affect penetration of the merozoite (Duffy blood group); (2) other factors can impede the intracellular development of the parasite (G6PD); or (3) parasitized cells can be more readily removed from the circulation (HbS).

The literature on blood groups and disease has been thoroughly reviewed by Mourant et al. (1978) and only a few examples will be given here. Blood group associations with infectious and non-infectious diseases have been reported. Blood group O may be related to resistance against smallpox and susceptibility to cholera. Group O hosts were favoured for blood meals by mosquito vectors for yellow fever. Group A persons have a lower incidence of scarlet fever, whereas group O people may be more susceptible to rheumatic fever. Also of interest is the finding that the number of secretors having rheumatic fever and carrying the type A strain, which is responsible for scarlet fever, was fewer than expected.

The first unequivocal association between blood group and disease was that between group A and gastric carcinoma (Aird et al. 1953). In general, the neoplasias show an association of malignancy with group A and to a lesser extent with group B. Secretor status associates with salivary gland cancer (excess of secretors). The evolutionary significance of the latter disease associations is unclear as they tend to be of adult onset and unlikely to affect reproductive fitness.

Haemolytic disease of the newborn results from the normal immune response of the mother to an incompatible group in her unborn child (Bias

1981), for instance a rhesus-positive fetus carried by a rhesus-negative mother. At parturition it is common for fetal blood to cross the placenta into the maternal circulation. This event, at the end of the first incompatible pregnancy, is usually the primary immunological challenge. The quantity of red blood cells that leaks across the placenta during the first pregnancy causes haemolytic disease in only 10 per cent of babies. This rises to 30 per cent for a second pregnancy, and thus it is third and subsequent pregnancies that are most at risk.

The pathology occurs when the antibody produced by the mother diffuses across the placenta into the fetal circulation where it destroys fetal red blood cells. As the fetal liver is incapable of removing the accumulated waste products the baby is born jaundiced or at worst severely ill or stillborn.

For an Rh-negative mother to have an Rh-positive fetus, the father has to be Rh-positive. If the father is homozygous all children of the mating will be Rh-positive, but heterozygous. If the father is heterozygous, half the children will be expected to be Rh-positive and half Rh-negative. The maternal–fetal incompatibility results in selection against heterozygotes. This would be expected to result in a transient polymorphism with the elimination of the less common allele.

However, fixation of Rh allele frequencies has not occurred although Rh-positive frequencies are usually higher than Rh-negative. Reproductive compensation has been postulated as a possible counteracting mechanism. This can occur where the father is heterozygous. The loss of Rh-positive children would be replaced by Rh-negative so as to produce the desired amount of offspring. Thus, the population frequency of Rh-negative alleles would increase.

Disease has also been associated with the human leucocyte antigen (HLA) histocompatibility system. For instance HLA-linked genes are involved in the susceptibility of the host to 'tuberculoid' (or localized lesion) form of leprosy and to tuberculosis. The majority of the autoimmune diseases show an HLA relationship. HLA associations with the outcome of exposure to HIV virus have recently been reported. A strong association was found between the HLA haplotype A1 B8 DR3 and development of HIV-related symptoms. Recently, scientists in Britain and The Gambia have discovered two HLA genes that seem to protect people against severe malaria.

Major geographic, urban–rural, and ethnic differences exist in the prevalence and incidence of diabetes. There are two main types of diabetes, type I and type II. Both are characterized by high blood sugar and they also both produce chronic vascular problems as a secondary complication. The genetic factors which contribute to each type appear to be quite distinct, as are the pathophysiological processes which lead to the development of the disease. Type I or insulin dependent diabetic (IDDM) does not produce any insulin and therefore cannot survive long without insulin replacement. Type

II or non-insulin dependent diabetic (NIDDM) produces insulin but for a variety of reasons cannot utilize it efficiently. Type II diabetics often have no symptoms or relatively mild symptoms, such as increased thirst, and thus this type of diabetes can go undiscovered for many years during which chronic vascular complications may develop.

IDDM is relatively uncommon in India, in the black population of sub-Saharan Africa, in the Japanese, and indeed in most 'non-Caucasians' (Hutt and Burkitt 1986). However, the question of whether IDDM is rare in developing countries still remains unanswered as there is a probability of early death without diagnosis of diabetes and this would account for the low rates which are mostly based on hospital data.

NIDDM is much the commoner form of diabetes and this type affects about 1 in 20 Americans and has been increasing to epidemic proportions among many Native Americans, Mexican-Americans, and migrants from India. It has been suggested that this type of diabetes represents a genetically determined adaptation to changes in life style which Neel (1962) accounted for by his 'thrifty genotype' hypothesis. According to this model, a thrifty gene provides protection against starvation in regions of great seasonal fluctuations of food supply by allowing its carriers to accumulate more fat than non-carriers. The one-time advantage has now become a mal-adaptation to the new environment of present-day food consumption patterns which, combined with a more sedentary life style, leads to diabetes. Although the physiological pathways proposed by Neel have been shown to be untenable, the genetic theory remains, and the hypothesis may well explain the high frequency of this condition in certain Amerindian groups of the American Southwest, such as the Pima and Papago (Neel 1971). However, as noted by Weiss and Chakraborty (1982): . . . 'it may be just a secondary effect of reservation existence. The fact that its present expression is as a degenerative disease does not imply that its selective effect in the past was expressed only late in life.'

Asians from the Indian subcontinent have received greater attention in diabetes studies because of their migration in large numbers before and after colonization. The prevalence of diabetes in migrant Indians was initially found to be higher than in the population residing in the Indian sub-continent and is usually higher than in the predominant ethnic group in the host country. Such findings have been reported from countries such as Singapore, Fiji, South Africa, Uganda, Trinidad, and the United Kingdom. The prevalence of diabetes is also higher in other migrant groups, such as Japanese in Hawaii, the Chinese and Malays in Singapore, the Yemenites in Israel, the West Indians in the United Kingdom, and the Tokelauans in New Zealand.

Migration may lead to an increase in the prevalence of NIDDM in a number of ethnic groups in parallel with social and cultural changes. The environmental factors may unmask NIDDM in a genetically susceptible individual. The disease appears to be associated with changing life style including increased longevity, dietary changes from traditional foods, and increased stress.

Family studies have shown a high prevalence of diabetes among offspring of conjugal NIDDM parents in India (Vishwanathan *et al.* 1985) and there is some evidence of a greater paternal influence in the transmission of NIDDM. Recent studies from South Africa and southern India have shown that NIDDM in Indian patients occurs at a younger age when compared with European populations. It may be that the genetic mechanisms are stronger in Indians. As NIDDM is the only type of diabetes in which simple autosomal inheritance has been implicated, it could be more frequent in populations where the disease is largely inherited. Alternatively, the younger age at diagnosis may be related to the younger age structure of the general population in these countries. A weak association between HLA antigens and NIDDM has been reported but its significance is small compared with that seen for IDDM (Eswhege *et al.* 1985). However, no associations have been reported between HLA antigens and NIDDM in Caucasoid populations (Zimmet *et al.* 1986).

Inbreeding is associated with an increased incidence of inborn errors (Chakraborty and Chakravarti 1977). Inbreeding is a feature among the Dravidians in southern India, with only a negligible percentage of marriages occurring between different castes. Consanguineous marriages have been occurring for many generations and one study found 47 per cent of marriages were between relatives. This has led some authors to suggest that inbreeding has contributed to the higher prevalence of diabetes in southern India and in the Tamil Indian community in South Africa (Jackson 1978), as well as in small island populations, such as Fiji, Malta, Mauritius, and Naura.

Diet may also contribute to the development of diabetes in two ways, quantitatively by supplying calories and if physical activity is low, by resultant obesity and qualitatively by the effects of specific foods. The interaction of diet, exercise, and obesity is complex and it is difficult to isolate their individual effects. In 1980, the second WHO Expert Committee on Diabetes concluded that the most powerful risk factor for NIDDM was obesity. Previous studies have shown both a strong and weak relationship with obesity. In the Western world, two-thirds or more of patients with diabetes are obese compared with NIDDM patients in India where obesity is uncommon. A possible reason is the confounding influence of weight loss with onset of the disease or with therapy.

Even so, Ahuja (1985) concluded that the difference in diabetes prevalence in Indians abroad could not be explained on the basis of adiposity when he compared Indians in India, Malaysia, and South Africa. Zimmet *et al.* (1983) was unable to explain the high prevalence of diabetes in the Fiji Indian population on the basis of Body Mass Index and Nicholl *et al.* (1986) could not attribute the markedly higher diabetes prevalence in British South Asians as compared with Europeans solely to the greater degree of adiposity.

Studies of identical twins suggest that the genetic component of NIDDM acts independently of obesity and therefore there may be additional environmental factors operating independently of obesity which contribute to a higher degree of diabetes prevalence in migrants and urban dwellers.

It would appear that the increase in prevalence of diabetes is a universal phenomenon of migration and Westernization. In Fiji, the increase in prevalence of diabetes was less over the years in migrant Indians than in natives. The prevalence increased 3.5 times in migrant Indians but 25 times in native Fijians in 15 years. However, there is little doubt that Asian Indians have a genetic susceptibility to develop NIDDM which becomes exposed when they migrate and/or achieve improved socio-economic status.

The present evidence suggests that tissue resistance to the action of insulin, giving rise to increased pancreatic secretion, may be a common feature in migrant Indians, leading to the high prevalence of NIDDM and other metabolic disturbances that are possibly responsible for the high rates of coronary heart disease.

CONCLUSIONS

In the space of a few decades we have seen the absolute eradication of one infectious disease, smallpox and the virtual elimination, particularly in the developed world, of many diseases, such as diphtheria and poliomyelitis, which were major killers.

However, human activities have and will continue to modify the patterns and biological impact of disease. As the environment changes and living and working conditions improve, new diseases emerge. For example, as air conditioning is increasingly used, the possibility of infection by bacteria which live in air conditioning units is increased, with *Legionella pneumophila* being the famous case. As personal and sexual behaviour changes, so does the threat from bacteria such as *Chlamydia* in causing urinary and pelvic infections. Crowding and other factors of urban life will maintain the transmission of respiratory and intestinal diseases and as human behaviour changes so will the behaviour of pathogenic organisms.

It is inevitable that humans will continue to discover that infection is at least one of the factors in inducing diseases for which no known cause exists, for instance many neurological conditions as well as many cancerous diseases of the blood.

How better to conclude but with words of René Dubos who wrote in *The mirage of health*: 'The very process of living is a continual interplay between the individual and his environment, often taking the form of a struggle resulting in injury or disease.'

REFERENCES

Ahuja, M. M. S. (1985). Heterogeneity in tropical diabetes mellitus. *Diabetologia*, 28, 708.

Aird, I., Bentall, H. H., and Fraser Roberts, J. A. (1953). A relationship between cancer of stomach and ABO blood groups. *British Medical Journal*, i, 799–801.

Allison, A. C. (1954). Protection by the sickle-cell trait against subtertian malarial infection. *British Medical Journal*, i, 290.

Allison, A. C. (1964). Population genetics of abnormal hemoglobins. In *Council for International Organization for Medical Science Symposium on abnormal haemoglobins*. Blackwell, Oxford.

Anderson, R. M. (1985). Mathematical models in the study of the epidemiology and control of ascariasis in man. In *Ascariasis and its public health significance* (ed. D. W. T. Crompton, M. C. Nesheim, and Z. S. Pawlowski), pp. 39–67. Taylor and Francis, London.

Armelagos, G. J., Ryan, M., and Leatherman, T. (1990). Evolution of infectious disease: a biocultural analysis of AIDS. *American Journal of Human Biology*, 2, 353–64.

Armstrong, B. and Doll, R. (1975). Environmental factors and cancer incidence and mortality in different countries, with special reference to dietary practices. *International Journal of Cancer*, 15, 617–31.

Awad el Karim *et al*. (1980). Quantitative egg excretion and work capacity in a gezira population infected with *S. mansoni*. *American Journal of Tropical Medicine and Hygiene*, 20, 54–61.

Barbosa, F. S., Coelho, M. U., and Dobbin, J. E. (1954). Qualidades de vetor dos hospederios de *Schistosoma mansoni* no nordeste do Bresil. II. Duracao de infestacao e eliminacao de cercarias em Australobis glabratus. *Publicaciones Avulses Instituto Aggeu Magalhaes*, 3, 78–93.

Belcher, D. W. *et al*. (1975). Guinea-worm in southern Ghana: its epidemiology and impact on agricultural productivity. *American Journal of Tropical Medicine and Hygiene*, 24, 243–9.

Belsey, M. (1976). The epidemiology of infertility: a review with particular reference to sub-saharan Africa. *Bulletin of the World Health Organization*, 54, 319–41.

Bias, W. B. (1981). Genetic polymorphism and human disease. In *Biocultural aspects of disease* (ed. H. E. Rothschild), Academic Press, New York.

Bjarnason, O., Day, N., Snaedal, G., and Tulinius, H. (1974). The effect of year of birth on the breast cancer age-incidence curve in Iceland. *International Journal of Cancer*, **18**, 689–96.

Bjelke, E. (1975). Dietary vitamin A and human lung cancer. *International Journal of Cancer*, **15**, 561–5.

Black, F. L. (1980). Modern isolated pre-agricultural populations as a source of information on prehistoric epidemic patterns. In *Changing disease patterns and human behaviour* (ed. N. F. Stanley and R. A. Joske), pp. 37–54. Academic Press, London.

Bradley, D. J., Rahmathullah, and Narayan, R. (1988). The tea plantation as a research ecosystem. In *Capacity for work in the tropics* (ed. K. J. Collins and D. F. Roberts), Cambridge University Press.

Brown, K. H. and Gilman, R. (1986). Nutritional effects of intestinal helminths with special reference to ascariasis and strongyloidiasis. In *The interaction of parasitic diseases and nutrition* (ed. C. Chagas and G. T. Keusch), pp. 213–32. Pontifica Academia Scientiarum, Vatican City.

Bruce-Chwatt, L. J. (1980). *Essential malariology*. Heinemann Medical Books, London.

Bruyning, C. F. A. (1985). Epidemiology of gastrointestinal helminths in human populations. In *Chemotherapy of gastrointestinal helminths* (ed. H. Van den Bossche, D. Thienpont, and P. G. Janssens). Springer, Berlin.

Buell, P. (1973). Changing incidence of breast cancer in Japanese-American women. *Japan N.C.I.*, **51**, 1479–83.

Bundy, D. A. P. (1986). Epidemiological aspects of Trichuria and trichuriasis in Caribbean communities. *Transactions of the Royal Society of Tropical Medicine and Hygiene*, **80**, 706–18.

Carrera, E., Nesheim, M. C., and Crompton, D. W. T. (1984). Lactose mal-digestion in Ascaris-infected preschool children. *American Journal of Clinical Nutrition*, **39**, 255–64.

Cerf, B. J., Rhode, J. E., and Sosanto, T. (1981). Ascaris and malnutrition in a Balinese village: a conditional relationship. *Tropical and Geographical Medicine*, **33**, 367–73.

Chakraborty, R. and Chakravarti, A. (1977). On consanguineous marriages and the genetic load. *Human Genetics*, **36**, 47–54.

Cockburn, T. A. (1967a). Infections of the order Primates. In *Infectious diseases: their evolution and eradication* (ed. T. A. Cockburn). Charles C. Thomas, Springfield, IL.

Cockburn, T. A. (1967b). The evolution of human infectious diseases. In *Infectious diseases: their evolution and eradication* (ed. T. A. Cockburn). Charles C. Thomas, Springfield, IL.

Collins, K. C., Brotherhood, J. R., Davies, C. T. M., Dore, C., Hackett, A. J., Imms, F. J. *et al.* (1976). Physiological performance and work capacity of Sudanese cane cutters with *Schistosoma mansoni* infection. *American Journal of Tropical Medicine and Hygiene*, **25**, 410–21.

Doll, R. (1955). Mortality from lung cancer in asbestos workers. *British Journal of Industrial Medicine*, **12**, 81.

Dubos, R. *The mirage of health: utopias, progress and biological change*. Harper and Row, New York.

Dunn, F. L. and Janes, C. R. (1986). Medical anthropology and epidemiology. In *Anthropology and epidemiology* (ed. C. R. Janes, R. Stall, and S. M. Gifford). Reidel, Boston.

Eswhege, E., Ducimetiere, P., and Thibult, N. (1985). Coronary heart disease mortality in relation with diabetes, blood glucose and plasma insulin levels. The Paris prospective study ten years later. *Hormone Metabolism Research*, **15**, 41–5.

Farooq, M. and Mullah, M. B. (1966). The behavioral pattern of social and religious water-contact activities in the Egypt-49 bilharziasis project area. *Bulletin of the World Health Organization*, **35**, 377–87.

Faust, E. C. and Russell, P. F. (1964). *Craig and Faust's clinical parasitology*. Lea and Febiger, Philadelphia.

Fenner, F. (1980). Sociocultural change and environmental diseases. In *Changing disease patterns and human behaviour* (ed. N. F. Stanley and R. A. Joske), pp. 7–26. Academic Press, London.

Foo, L. C. (1986). Impact of *Ascariasis lumbricoides* and *Trichuris trichuria* on growth of early school age Tamil–Malaysian children. D.Phil. Thesis, Cornell University, Ithaca, New York.

Forsum, E., Nesheim, M. C., and Crompton, D. W. T. (1981). Nutritional aspects of *Ascaris* infection in young protein-deficient pigs. *Parasitology*, **83**, 497–512.

Frisch, R. E. (1984). Body fat, puberty and fertility. *Biological Reviews*, **59**, 161–88.

Frisch, R. E. and Revelle, R. (1971). Height and weight at menarche and a hypothesis of menarche. *Archives of Diseases in Childhood*, **46**, 695–701.

Gajdusek, D. C. (1973). Kuru in the New Guinea Highlands. In *Tropical neurology* (ed. J. D. Spillane). Oxford University Press.

Gregory, K. F., Carpenter, J. A., and Bending, G. C. (1967). Infection hazards of the common communion cup. *Canadian Journal of Public Health*, **58**, 305–10.

Gwatkin, D. R. (1984). Does better health produce greater wealth? A review of the evidence concerning health, nutrition and output. Unpublished USAID document.

Haldane, J. B. S. (1956). *The causes of evolution*. Cornell University Press, Ithaca.

Halloran, M. E., Bundy, D. A. P., and Pollitt, E. (1989). Infectious disease and the UNESCO Basic Education Initiative. *Parasitology Today*, **5**, 359–62.

Hill, D. L. and Grubbs, C. J. (1982). Retinoids as chemopreventive and anticancer agent. *Anticancer Research*, **2**, 111–24.

Hutt, M. S. R. and Burkitt, D. P. (1986). *The geography of non-infectious disease*. Oxford University Press.

Hirayama, T. (1979). Diet and cancer. *Nutrition and Cancer*, **1**, 67–81.

Holland, C. (1987). Neglected diseases—trichuriasis and strongyloidiasis. In *The impact of helminth infections on human nutrition—schistosomes and soil transmitted helminths* (ed. L. S. Stephenson), pp. 161–201. Taylor and Francis, London.

Jackson, W. P. U. (1978). Epidemiology of diabetes in South Africa. *Advances in Metabolic Disorders*, **9**, 111–15.

Kaplan, B. A. (1988). Migration and disease. In *Biological aspects of human migration* (ed. C. G. N. Mascie-Taylor and G. W. Lasker). Cambridge University Press.

Khalil, M. (1949). The national campaign for the treatment and control of schisto-somiasis from the scientific and economic aspects. *Journal of the Royal Egyptian Medical Association*, **32**, 817–56.

Kinlen, L. J. (1982). Meat and fat consumption and cancer mortality: a study of strict religious orders in Britain. *Lancet*, **1**, 946–9.

Kvalsvig, J. D. (1988). The effects of parasitic infection on cognitive performance. *Parasitology Today*, **2**, 80–1.

Kwast (1989). Maternal mortality: levels, causes and promising interventions. *Journal of Biosocial Science Supplement*, **10**, 51–67.

Layrisse, M. and Roche, M. (1964). The relationship between anemia and hook-worm infection: results of surveys of rural Venezuelan population. *American Journal of Hygiene*, **79**, 279–301.

Livingstone, F. B. (1957). Sickling and malaria. *British Medical Journal*, i, 762.

Livingstone, F. B. (1958). Anthropological implication of sickle-cell gene distribu-tion in West Africa. *American Anthropologist*, **60**, 533–52.

Livingstone, F. B. (1971). Malaria and human polymorphism. *Annual Review of Genetics*, **5**, 33–64.

Livingstone, F. B. (1984). The Duffy blood groups, vivax malaria and malaria selection in human populations: a review. *Human Biology*, **56**, 413–25.

Long-de, W. and Hammond, E. C. (1985). Lung cancer, fruit, green salad and vitamin pills. *Chinese Medical Journal*, **3**, 206–10.

Lunn, P. G. (1988). Malnutrition and fertility. In *Natural human fertility and bio-logical determinants* (ed. P. Diggory, M. Potts, and S. Teper). Macmillan Press in association with The Eugenics Society, London.

Macdonald, G. (1973). *Dynamics of tropical disease*. Oxford University Press.

Mahalanabis, D., Simpson, T. W., and Chakraborty, M. L. (1979). Malabsorption of water miscible vitamin A in children with giardiasis and ascariasis. *American Journal of Clinical Nutrition*, **32**, 313–18.

Mancuso, T. F. and Coulter, E. J. (1964). Methodology in industrial health studies. *Archives of Environmental Health*, **6**, 210–26.

Marmot, M. G. (1980). Affluence, urbanization and coronary heart disease. In *Disease and urbanization* (ed. E. J. Clegg and J. P. Garlick). Taylor and Francis, London.

Martorell, R. and Gonzalez-Cossio, T. (1987). Maternal nutrition and birth weight. *Yearbook of Physical Anthropology*, **30**, 195–216.

McFalls, J. A. and McFalls, J. H. (1984). *Disease and fertility*. Academic Press, Orlando.

McNeill, W. H. (1976). *Plagues and peoples*. Doubleday, London.

Miller, L. H. (1977). Current prospects and problems for a malaria vaccine. *Journal of Infectious Diseases*, **135**, 858–64.

Mosley, W. H. (1979). The effects of nutrition on natural fertility. In *Patterns and determinants of natural fertility* (ed. J. A. Menken and H. Leridon), pp. 83–105. Ordina, Leige.

Mourant, A. E., Kopec, A. C., and Domaniewska-Sobczak (1978). *Blood groups and diseases*. Oxford University Press.

Neel, J. V. (1962). Diabetes mellitus: a 'thrifty' genotype rendered detrimental by 'progress'? *American Journal of Human Genetics*, **14**, 353–62.

Neel, J. V. (1971). Genetic aspects of the ecology of disease in the American Indian. In *The ongoing evolution of Latin America populations* (ed. F. Salzano). Charles C. Thomas, Springfield, IL.

Nicholl, C. G., Levy, J. C., Mohan, V., Rao, P. V., and Mather, H. M. (1986). Asian diabetics in Britain. A clinical profile. *Diabetic Medicine*, 3, 257–60.

Nwosu, A. B. C., Ifezulike, E. O., and Anya, A. O. (1982). Endemic dracontiasis in Anambra State of Nigeria: geographical distribution, clinical features, epidemiology and socio-economic impact of the disease. *Annals of Tropical Medicine and Parasitology*, 76, 187–200.

Omer, A. H. S. and Ahmed, El Din (1974). Assessment of physical performance and lung function in *Schistosoma mansoni* infection. *East African Medical Journal*, 51, 217–22.

Parker, M. (1989). The effects of *Schistosoma mansoni* on female activity patterns and infant growth in Gezira Province, Sudan. Unpublished D.Phil. thesis, University of Oxford.

Parry, E. H. O. (1984). *Principles of medicine in Africa* (2nd edn). Oxford University Press.

Phillips, R. L., Garfinkel, L., Kuzma, J. W., Beeson, W. L., Lotz, T., and Brin, B. (1980). Mortality among Californian seventh-day adventists. *Cancer Research*, 35, 3513–22.

Pike, M. C. (1988). Fertility and its effects of health. In *Natural human fertility and biological determinants* (ed. P. Diggory, M. Potts, and S. Teper). Macmillan Press in association with The Eugenics Society.

Polgar, S. (1964). Evolution and the ills of mankind. In *Horizons of anthropology* (ed. S. Tax). Aldine, Chicago.

Prescott, N. M. (1979). Schistosomiasis and development. *World Development*, 7, 1–14.

Prior, I. A., Stanhope, J. M., Evans, J. G., and Salmond, C. E. (1974). The Tokelau Island migrant study. *International Journal of Epidemiology*, 3, 939–53.

Reynolds, V. and Tanner, R. E. S. (1983). *The biology of religion*. Longman, London.

Robinson, D. (ed.) (1985). *Epidemiology and the community control of disease in warm climate countries*. Churchill Livingstone, Edinburgh.

Roche, M. and Layrisse, M. (1966). The nature and causes of hookworm anaemia. *American Journal of Tropical Medicine and Hygiene*, 15, 1030–1100.

Rucknagel, D. L. and Neel, J. V. (1961). The hemoglobinopathies. *Progress in Medical Genetics*, i, 158.

Russel-Briefel, R., Bates, M. W., and Kuller, L. H. (1985). The relationship of plasma carotenoids to health and biochemical factors in middle-aged men. *American Journal of Epidemiology*, 122, 741–9.

Sattenspiel, L. and Castillo-Chavez, D. (1990). Environmental context, social interactions, and the spread of HIV. *American Journal of Human Biology*, 2, 397–418.

Schofield, C. (1985). Parasitology today: an ambitious project. *Parasitology Today*, 1, 2.

Selikoff, I. J., Chung, J., and Hammond, E. C. (1964). Asbestos exposure and insulation workers in the United States and Canada. *Annals of the New York Academy of Science*, **330**, 91–116.

Shekelle, R. B., Lepper, M., and Liu (1981). Dietary vitamin A and risk of cancer in the Western Electric Study. *Lancet*, **2**, 1186–90.

Sivakumar, B. and Reddy, V. (1975). Absorption of vitamin A in children with ascariasis. *Journal of Tropical Medicine and Hygiene*, **78**, 114–15.

Sporn, M. B. and Roberts, A. B. (1983). Role of retinoids in differentiation and carcinogenesis. *Cancer Research*, **43**, 3034–40.

Stanley, N. F. (1980). Man's role in changing patterns of arbovirus infections. In *Changing disease patterns and human behaviour* (ed. N. F. Stanley and R. A. Joske), pp. 151–74. Academic Press, London.

Stein, Z. and Susser, M. (1978). Famine and fertility. In *Nutrition and human reproduction* (ed. W. H. Mosley), pp. 11–28. Plenum, New York.

Stephenson, L. S. (1987). *The impact of helminth infections on human nutrition—schistosomes and soil transmitted helminths*, pp. 161–201. Taylor and Francis, London.

Stephenson, L. S., Latham, M. C., Kurz, K. M., Miller, D., Kinoti, S. N., and Oduori, M. L. (1985). Urinary iron loss and physical fitness of Kenyan children with urinary schistosomiasis. *American Journal of Tropical Medicine and Hygiene*, **34**, 519–28.

Stephenson, L. S., Latham, M. C., and Mlingi, B. A. (1986). Schistosomiasis and human nutrition. In *Schistosomiasis and malnutrition* (ed. L. S. Stephenson), Cornell International Nutrition Monograph Series No. 16, Cornell International Nutrition Program, Ithaca, New York.

Stoll, N. R. (1947). This wormy world. *Journal of Parasitology*, **33**, 1–18.

Stoll, N. R. (1962). Helminthic infections. In *Drugs, parasites and hosts*. Biological Council Symposium (ed. L. G. Goodman and R. H. Nimmo-Smith). Churchill-Livingstone, London.

Stryker, W. S., Kaplan, L. A., Stein, E. A., Stampfer, M. J., Sober, A., and Willett, W. C. (1988). The relation of diet, cigarette smoking, and alcohol consumption to plasma beta-carotene and alpha-tocopherol levels. *American Journal of Epidemiology*, **127**, 283–96.

Tannenbaum, A. and Silverstone, H. (1953). Nutrition in relation to cancer. *Advances in Cancer Research*, **1**, 451–501.

Taren, D. L., Nesheim, M. C., and Crompton, D. W. T. (1987). Contributions to poor nutritional status in children from Chirqui Province, Republic of Panama. *Parasitology*, **95**, 615–22.

Tripathy, K., Duque, E., Bolanos, O., Lotero, H., and Mayoral, L. G. (1972). Malabsorption syndrome in ascariasis. *American Journal of Clinical Nutrition*, **25**, 1276–87.

Underwood, P. and Underwood, Z. (1980). Expectations and realities of western medicine in a remote tribal society in Yemen, Arabia. In *Changing disease patterns and human behaviour* (ed. N. F. Stanley and R. A. Joske). Academic Press, London.

Viswanathan, M., Mohan, V., and Snehalatha, C. (1985). High prevalence of diabetes among offspring of conjugal type 2 diabetic parents in India. *Diabetologia*, **28**, 907–10.

Weiner, I., Burjke, L., and Godberger, M. A. (1951). Carcinoma of the cervix in Jewish women. *American Journal of Obstetrics and Gynecology*, **61**, 418–22.

Weisbrod, B. A., Andreano, R. L., Baldwin, R. E., Epstein, E. H., and Kelley, A. C. (1973). *Disease and economic development. The impact of parasitic diseases in St Lucia.* University of Madison Press, Madison.

Weiss, K. M. and Chakraborty, R. (1982). Genes, polymorphisms and disease. In *A history of American physical anthropology, 1930–1980* (ed. F. Spencer). Academic Press, New York.

Willett, W. (1990). *Nutritional epidemiology.* Oxford University Press.

Wood, W. B. and Gloyne, S. R. (1934). Pulmonary asbestosis: a review of one hundred cases. *Lancet*, **2**, 1383.

(WHO) World Health Organization (1979). *International classification of diseases, injuries, and causes of death*, WHO, Geneva.

(WHO) World Health Organization (1984a). Schistosomiasis—new goals. *World Health*, December 1984.

(WHO) World Health Organization (1984b). Intestinal parasitic infections and how to prevent them. *World Health*, March 1984.

Wright, W. H. (1951). Medical parasitology in a changing world. What of the future? *Journal of Parasitology*, **37**, 1–12.

Zimmet, P., Taylor, R., and Ram, P. (1983). Prevalence of diabetes and impaired glucose tolerance in the biracial (Melanesian and Indian) population of Fiji: a rural urban comparison. *American Journal of Epidemiology*, **118**, 673–88.

Zimmet, P., Serjeantson, S. W., King, H., and Kirk, R. (1986). The genetics of diabetes mellitus. *Medical Journal of Australia and New Zealand Journal of Medicine*, **16**, 419–24.

Zuluata, J. de (1956). Malaria in Sarawak and Brunei. *Bulletin of The World Health Organization*, **15**, 651–75.

2

SOME STUDIES OF SOCIAL CAUSES OF AND CULTURAL RESPONSE TO DISEASE

G. Lewis

INTRODUCTION

The premiss of this book is an obvious truth—that disease depends on both biological and social facts. It may be obvious but it can be extremely hard to unravel the exact connections. The term 'disease' covers a multitude of conditions. Malaria, multiple sclerosis, hernia, and heart failure have little in common except that they are all examples of what doctors may be expected to deal with as medical matters. Ideas about illness are diverse and there are many kinds of disease.

The point about diversity applies to society as well and hence to the social factor in disease. If there are many kinds of disease with varied characteristics, there are also many kinds of society. They vary in scale and organization, internal division, and complexity. In some parts of New Guinea a village-sized community may once have acted as a sovereign political unit and been an independent society. In social rules, language, territory, shared experience, and integration, it may still be well demarcated and distinct. But political, cultural, and social boundaries do not necessarily correspond neatly. Specific customs or rules, which may be relevant to distributions of disease, need not match political boundaries. Many modern states are very large, e.g. India and the United States, and they pose obvious problems for identifying 'the society', or the specific cultural environment or class to which someone belongs, the sense in which it is distinct, homogeneous, cohesive, or clearly bounded in social or cultural terms. The examples to follow will serve to illustrate some interrelations between disease and society.

The complexity of cause

In 1960, Charles Wilcocks introduced his Heath Clark lectures on *Aspects of medical investigation in Africa* by remarking that study of the relationships between people and their environment must include 'the beliefs, habits and fears of people as well as the physical surroundings in which they live' (Wilcocks 1962, p. xii). It was not common for medical scientists to give a prominent place to social and cultural factors in explaining patterns of disease. However, the view he developed in his lectures supports the idea that a social anthropological approach might be useful in medical investigations.

For his first example of the investigation of disease in Africa, Wilcocks chose not a characteristic 'tropical disease' but tuberculosis (TB). During World War I, high rates of illness and mortality from TB were observed among Senegalese soldiers who had been brought to Europe. They were thought to lack innate resistance, to be in effect 'virgin soil' for the disease. The type of disease they showed was often severe, with marked fever, pneumonia, emaciation, generalization of the infection—some had miliary spread of TB, i.e. infection widespread in many tissues; it was overwhelming, and they died rapidly. They showed little tendency to the healing and fibrosis found in European patients.

It was easy to suppose that TB was new to them, that as members of ethnic groups which had not before encountered the disease, the bacillus had never had an opportunity to act selectively on their populations. Hence their susceptibility, the severity of the effects, and the sense in which the Senegalese soldiers could be seen as 'virgin soil' for TB.

Another suggestion to explain their response to infection was that the Africans taken out of their familiar environments, subjected to the hardships of military life in strange conditions, were exposed both to infection and to considerable stress; the stress made them more vulnerable to infection. Other studies many years later in South Africa showed that, of the African mineworkers who developed TB at the mines, many had already encountered the TB bacillus before at their homes, had reacted to it and remained well, but under the combined stresses of migration, bad housing, the work conditions in the mines, quiescent or healed primary foci of previous infection broke down (Wilcocks 1962, chapter 1).

The issues in this example suggest how complex a full answer would need to be if one were to try to pursue all the social and biological facts that might contribute to explain a particular pattern of disease in some population. Exposure, susceptibility and reactions to disease would involve questions about the bacteria, modes of exposure to infection, living and working conditions, diet and income, the politics and history of contact, the physiology and psychology of stress, the people's perceptions of their situation, their culture, their understanding of illness, and their behaviour in response to it.

Levels of inquiry

Medical questions may thus require investigation at the level of the environment and population, at that of the cell and the molecule, or at any level in between. The focus may be narrowed to increasingly precise detail: from the individual patient, to physiological mechanisms, a particular organ, right down to cellular and biochemical levels. That, of course, has been the traditional medical approach. But we could take instead a social or an epidemiological perspective—the questions widening outwards from observations on the people who are ill, to the conditions in which they live, their social and family histories, the biographical and personal events surrounding their illness, how they became ill, what they did in response to it, and the reasons for their actions. How fruitful or relevant some of these questions might turn out to be would, of course, depend on the subject.

The approaches are complementary—detailed knowledge about internal mechanisms may sometimes serve to identify more precisely other questions to be explored at the level of social behaviour or situation or environment.

To show the meaning of this interplay between a narrow and a broad approach, I will discuss an example in which interrelationships between social and biological factors in disease have been studied in great detail. The example comes from the work of the Medical Research Council's Dunn Nutrition Unit research station at Keneba in The Gambia, West Africa. It is intended to show how environment, subsistence, and social behaviour are intricately bound up with the explanation of some forms of disease: the studies to be described focus on illness during early childhood.

DISEASE IN A VILLAGE

Keneba, The Gambia

The origin of these studies in The Gambia goes back to 1949 when Dr (later Sir) Ian McGregor chose for long-term study four villages (including Keneba) in the West Kiang Administrative District. The district is an area of savannah scrub and farm land, roughly 40 × 20 kilometres in area, bounded on three sides by brackish tidal rivers, the Gambia, and the Bintang Bolong. The district now has a population of about 9000 people. McGregor chose the area and the particular villages because they were in what was then regarded as an isolated and backward district of The Gambia. The aim was to initiate research on the importance of endemic disease and its impact on health and nutritional status. The Medical Research Council supported these studies and the Unit, which in 1974 took over and developed the field research centre which McGregor had established in one of the villages—Keneba. Since 1974, the Dunn Nutrition Unit (DNU) has supported a wide variety of inquiries (Weaver and Campbell 1989; Whitehead 1985, 1988): these provide the data I am using here.

In McGregor's plan, clinical epidemiology was to be a prominent element of the work, especially study of the relationships between malnutrition and communicable disease. The four villages were all subsistence farming communities (growing a mixture of staples: rice, millet, sorghum, groundnuts) and, even by Gambian standards, they were poor. His intention was to document the epidemiology of whole communities in a rural environment. McGregor stressed the need to study them longitudinally, and to characterize the community and its mode of life. The resulting findings showed that malnutrition in African children was the result of more than just dietary inadequacy. They showed how epidemiological studies in the home environment could define and elucidate complex health problems (McGregor 1990). As a result of the studies, various measures to prevent or treat illness and promote health were later introduced.

One of the four villages refused after the first few years to stay in the study. The village leader at the time had tried to persuade them to stay in; in 1989 his nephew told me that it was because the people in the 1950s had disliked having the examinations and blood tests. They chose not to continue to take part in the surveys because quite a few of them had come to be suspicious of what was being done with their blood; they also thought that the pills or the linctus they were prescribed might contain alcohol: they were Muslim as were the people in all the villages.

The three villages remaining in the study comprised mainly of Mandinka-speaking people together with some Fula and Jola. These three villages have provided the subjects for regular census and survey over the following years. All inhabitants were recorded and examined annually, with investigations done at other times to cover seasonal variations and a wide variety of topics (there are data on morbidity, family structure, births and deaths, and the results of many clinical, laboratory, and social investigations, for example, on breast-milk intake, energy expenditure in different activities, and blood calcium levels). The survey villages now have immediate local access to a very high standard of primary health care provided by the DNU staff, headed at the research station in Keneba by a physician with specialist paediatric training.

The studies have continued for more than 30 years. They provide a record of the changing patterns of growth and survival. They soon showed the vulnerability of life in these villages—McGregor's early results indicated that one out of every two babies born alive would fail to survive to the age of 5 years. They revealed the marked changes in death- and illness-rates that came with the annual change from dry to wet season, the alternation between a period with adequate food supplies and a critical time of hunger and exhausting work. The farming cycle, set by climate and season, brought a time of greater hardship, food scarcity, and illness each year, while social rules about men's and women's tasks placed women with young children under especially heavy work burdens at the critical time of the year. Domestic duties, large families, complex arrangements for housing and feeding in big compounds, water supplies, food storage, disposal of excrement, are all directly and indirectly relevant to the patterns of disease experienced by the people; they are also all topics on which the villagers have their own opinions and feelings, and these affect their actions.

Social factors influencing illness in early life in The Gambia

On the question of health in infancy and early childhood and specifically diarrhoeal illness, one path of inquiry is to narrow the questions down. Which bacteria are involved? What effects on the small intestine are produced by recurrent infection? What is the extent of cellular and physiological damage to the intestinal lining? Is there evidence of malabsorption, malnutrition and longer-term growth faltering? This direction of inquiry reveals mainly the effects of disease.

Another path is to study infection: when it occurs, what factors bring it about—and this involves broad questions about the environment, the pollution of water, food contamination, the mother's work and her child-care duties. These questions ramify out to broader social issues: from, for example, weaning patterns and the reasons why mothers need to leave food aside for their children, to polygamy and the demands on women, to wider social, economic and political issues that affect people's choices and values, issues such as education, cash crop development and migration. The farming work expected of Gambian village women limits the amount of time and attention they can give to their young children. The women's work patterns contribute to explanations for the patterns of childhood illness. In contrast to the first direction of inquiry, the second one tends to be concerned with the causes of illness rather than effects of disease.

The sections which follow trace some of the causes of childhood illness, especially diarrhoea in early childhood. The care mothers can give small children is crucial and one needs to know about their mothers' daily lives and circumstances.

It is easy to accept that rainfall and the environment are part of a set of conditions which determine some aspects of infant and childhood illness.

The climate has effects on food supply, insect vectors of disease, work, water, and pollution. The social arrangements and the values that underlie family and economic life are contributory causes. Each year the villagers experience a return of scarcity and hunger in the face of exacting demands for long hours of work. Their loss of weight during the season of hardest work gives striking evidence of this. The villagers' preoccupation with the insecurity of food supplies is understandable.

Food and work pressures

Subsistence needs and perceptions of the threat of hunger lie behind many aspects of social organization. The alternation of wet and dry seasons imposes its rhythm on the farming year. The growing season in The Gambia is short, about five months, dictating the critical timing, key tasks and bottlenecks in demand for labour; each kind of crop has its own characteristics and needs. But the division of tasks and responsibilities in farming and family life does not follow automatically from the characteristics of the climate or the crop. Things may be done differently in other societies in similar settings. The organization of subsistence is bound up also with aspects of control and authority in social life and with cultural values. These are behind the division of labour and the different expectations and pressures put on men and women. They contribute therefore directly and indirectly to problems of coping and patterns of risk and illness.

Work pressures on mothers of small children have implications for their children's health. Village subsistence is dependent on rainfall; the quantity and timing have a marked effect on the size of the eventual harvest. So the rains dictate the agricultural calendar and the times of critical demand for work. The heaviest work (clearing ground and planting) comes in June, July, and August when food grain stocks are getting low or close to exhausted, and there is little cash left. As the rains reach their peak, women have to get the work done. They work harder and longer hours than men; for women this may mean a 15 hour working day with 10 to 12 hours spent away from home at the fields (Roberts *et al.* 1982).

Women are overworked, underfed, and underweight. In the wet season, their food intake averaged 1302 kcals per day and they experienced a sharp loss of body weight—a loss on average of 3.4 kg between July and October, and this had to be regained during the dry season. A lactating mother who is also working in the fields needs about 2800 kcals per day (see the table in Rowland *et al.* 1981, p. 172; even pregnant women lose weight during the rains). Gambian food is high in water and fibre content and very low in fat; it supplies roughly 1 kcal per g—bulk rather than energy. So the lactating mother who is also working in the fields would need to eat about 2.8 kg of food to meet her requirements: to eat such a daily bulk is almost impossible. And many women are lactating as most go from pregnancy to pregnancy.

Fertility

High value is set on fertility. The physical costs of having many children fall on the mothers. But the bigger the family, the more hands there will be to help with work—eventually. This outlook has been associated with a pattern of hoe agriculture typical of sub-Saharan Africa (Goody 1976) in which it is not so much lack of land as lack of labour that limits productivity. Access to labour and control over it gave advantage and relative security. The plough and the use of animal traction is a recent introduction in The Gambia. Traditional agriculture depended on human labour and the people to do it: high fertility, polygyny, large extended families provided them (in the past slaves were added in some families). Husbands expected their wives to supply domestic services, agricultural labour, subsistence foods, and to bear them children. Children are desired but their survival into adult life is uncertain.

The associated cultural values are deeply embedded. The average Keneba mother has had six or more full-term pregnancies by the time she reaches 35 years of age. Many by the end of their reproductive lives will have had between 10 and 14 pregnancies and produced between 3000 and 4000 litres of milk (Prentice and Prentice 1988). The burden on women in The Gambia is a double burden in the sense of being reproductive and productive. In addition to all their farming and domestic tasks, the women have to cope with pregnancy and lactation. The costs of being prolific fall directly on them.

Division of labour

Changes have occurred in the distribution of work between men and women, but in general they have not eased the problems for women of caring for young children. There are customary rules about rights and responsibilities in economic and domestic life. Men and women are both expected to contribute to the family food needs—the women producing rice, the men millet and groundnuts. But while women work on their husbands' millet or groundnuts, men do not work with their wives in the rice fields.

The work loads and costs are not equally distributed. A man, the family head, usually the woman's husband, controls and decides about the use of land; and his wife is expected to do as he says. Land and animals are predominantly owned by men, gardens and fields are dispersed. Women traditionally specialize in growing rice. In some cases, land for growing rice was owned by women individually (if they had cleared it or inherited it) and they could transmit it to their daughters. Women also tend various garden crops (lentils, cassava, eggplant, tomatoes, etc.) and some groundnuts; while the men farm the millet, sorghum, and groundnuts. Groundnuts have become the major (almost the only) cash crop of the Gambian economy; more land and labour is devoted to it.

The fields a family uses are scattered, partly because the areas suitable for wet rice farming are patchily dispersed and not close by, and partly to spread the risk of damage to crops by monkeys or wild pigs. Men's groundnut fields tend to be closer to the village, women's rice fields further away. This is one aspect of the relatively separate spheres of men and women in daily life. Another is commonly visible in the village during the day; when older men are to be seen relaxed and chatting in the shade, no women are present because the women are away in the fields.

Cash crop development

When new opportunities to earn cash were introduced, the men took advantage of them if they could, but quite often at the expense of the women on to whose shoulders shifted an increased burden of subsistence work: clearing, weeding, guarding crops, and harvesting them. Secondary schooling, employment, and the search for jobs took the men away from farming. Developments in farming tools and techniques, such as the introduction of ploughs and carts for transport, have been almost exclusively taken up and used by men. In other parts of The Gambia, men have tried to manipulate rules about land ownership and labour obligations to take advantage of new possibilities for irrigating land to produce rice (Carney 1988).

Women may earn money by hiring out their labour for work on groundnut farms other than ones belonging to their husbands, but they do not want to do this to the detriment of their immediate family subsistence needs. They can also earn money by selling produce from their vegetable gardens. A programme funded by Action Aid has provided wells in a large communal field on the outskirts of Keneba to help women produce vegetables. Its proximity is a boon as distance and lack of transport add appreciably to the daily burdens of farming.

Co-operation

In the evening, women cluster round the village taps fetching water; there are groups in the compounds of women pounding grain, sometimes two or three pounding together in the same big wooden mortar. The co-ordinated triple timing of three women pounding in the same mortar conveys an impression of co-operation in work by women which may also be evident in the housing and cooking arrangements of co-wives and in age-set work groups. Women belong to age sets (*kafo*) which are village-wide associations and cross the boundaries of lineage and family divisions. These *kafo* can be gathered for communal work on farms according to traditional rules of reciprocal exchange, or for pay; but their role is diminishing as families become more concerned with jobs and individual earning of cash.

Compounds and family

Keneba village has a reputation for strict Muslim observance. Men have authority over women and children, the old are treated with marked respect, and polygyny is the norm. With large extended families living together in one compound, domestic arrangements become complex (see Fig. 2.1). The size of a compound may range from a small one containing a single conjugal family unit to a large long-established compound containing perhaps over 120 people housed in many different houses and rooms, together with some hutches or huts for the sheep and goats, and a patch of garden for maize, tomatoes, etc., all within the main fence enclosing the compound.

The differences of scale, complexity, and internal arrangements are obviously great. Typically, a man would keep separate the houses or rooms for himself, his wife, a grain store, and a kitchen. The house or room for

Fig. 2.1 A typical Keneba compound.

his first wife will usually be shared if he marries additional wives. If co-wives get on happily together they continue to share the room and domestic duties. There are some advantages in sharing: they can alternate who does the cooking, if one is ill, etc. There are disadvantages: women and children sleep in markedly more crowded conditions then men and they are exposed to greater chances of cross-infection. One house or room may be shared by as many as four co-wives, each having her bed in one corner of the room. But quite commonly, as more children come, frictions between co-wives develop and they may wish to separate. But it is rare for a husband to provide his co-wives with different rooms or houses, although they may farm separate pieces of land, and choose to cook each at their own kitchens.

Consumption units and sharing

The effective domestic unit is the consumption unit (*sinkiro*) centred on a particular kitchen hearth. It is a cooking unit whose food and provisioning comes from the work of its members. The women are in control of daily arrangements. They plan the meals and get supplies; they do the cooking and serve the portions which go into the different bowls for the members of the *sinkiro*. Women share bowls with their young children. Men and grown up sons have separate bowls. Perhaps one contributory reason for men not appreciating the meagre rations on which women have to work in the hungry season is the men's relative advantage in having their own individual and more generous portions. Small children must compete for their share from a bowl; slow, weak or apathetic eaters, such as there might be from illness, are at an added disadvantage. One of the Gambian field survey workers who had been making extensive observations and weighing the detailed individual food intakes of a large number of families commented that women in general eat more slowly then men, talk and chat more while eating, and that they also hold back from taking food from the bowl to let their children get a bit more if they can.

A husband will receive food from each of his wives if each has a separate *sinkiro*. He should contribute to each wife's maintenance fairly. The old form of granary, a raised construction, has been replaced by a room with a lock on the door. Traditionally, the husband controlled the granary, but in many cases now the wife keeps the key to her grainstore and each wife may have her own. She may keep part of her earnings from selling garden produce, her labour, her rice. Despite appearances, the supposedly traditional pattern of patriarchal authority and control by the male head of the family may in fact leave much of the responsibility for decisions about daily food in the hands of the wife. There seems to be a growing move to fragment and individuate domestic arrangements (Beckerleg, forthcoming). The *sinkiro* pattern of organization had always been a flexible device to allow for the rearrangement of domestic relationships according to need and compatibility. It allowed for the inclusion of visitors or strangers, and for the addition of children for fostering. Consumption units could give each other temporary help, for instance, during illness. The flexible *sinkiro* arrangements make the detailed mapping of domestic relationships, both within the family and between families, sometimes very complex to analyse. The terms household or extended family may imply co-residence and the pooling or sharing of resources, but a large compound contains within it complex internal divisions and interrelationships.

Even without more information about village social structure, and wider issues, such as the effects of migration, jobs, or education, it should be clear that women's lives, work and responses to illness, are affected by these social arrangements and cultural values, as well as by political and economic changes in the wider world. Illness is not a matter solely of the biology of disease, climate, and natural environment. Social factors produce and can change some of the conditions in which mothers have to bring up their small children. The relationships between domestic and non-domestic labour shift, and may make it more difficult to find the time for nursing an infant, or put greater stress on a woman with many young children. Roads and transport, the advent of a bus service to the coast, the implantation of a large medical research station with all its associated staff, have had specific effects on health or welfare locally. The identification of precise effects would, no doubt, require one to follow many such paths. New tools, development projects, the price of groundnuts on the world market, the growth of tourism—all eventually produce local effects and some of them have implications for children's health, women's work, and patterns of exposure to illness. The ramifications are complex and often indirect. Child-care, child health, and household organization belong primarily in the domestic domain but they are connected in a multitude of ways with wider economic and political spheres.

ILL HEALTH IN EARLY CHILDHOOD

Clinical surveys have established the kinds of illness and how much of each has occurred. The worst months for child mortality and morbidity are August, September, and October (the second half of the rainy season which is also the hungry season), when nearly half of deaths in later infancy and early childhood occurred, with the peak age of child mortality in children aged between 9 and 14 months. McGregor *et al.* (1961) found a neonatal mortality rate (NNR) of 54 per 1000 live births and an infant mortality rate (IMR) of 134 per 1000 live births; in 1974, when continuous, free-of-charge, medical care was established, the neonatal mortality rate was still high (83.7), as was the infant mortality rate (148.8). By 1982–3, health and nutrition interventions had reduced mortality to NNR 45.5 and IMR 24.5 (Lamb *et al.* 1984). McGregor recorded the prevalence of various infectious diseases (including malaria, filariasis, hookworm, intestinal worms, tetanus, measles, whooping cough).

It was clear that diet and disease combined to produce the high vulnerability of early life. Recurrent illness caused severe faltering of growth. It depleted the child's scanty energy reserves in the acute phases of illness when infection increases metabolic expenditure; and there were longer-term consequences. In a child too ill to want to eat, lack of appropriate food slowed recovery, and apathy and weakness developed. Exposure to further infection was likely with the child in a weakened state. Malaria, respiratory tract infections, and diarrhoea were the main recurrent causes of illness showing marked seasonal variation (see the table in Rowland *et al.* 1981, p. 173). Malaria had been an early focus of McGregor's work. Malaria, including malignant malaria, was concentrated in the wet season, but Rowland *et al.* (1981) found that children suffered from diarrhoeal illness about 13 times as often as they did malaria. From the second half of a child's first year of life to the age of 3 years, many children experienced diarrhoea for roughly one week of each month of the rainy season. About 30 per cent of infants had a second attack of diarrhoea within a 28 day follow-up period (Weaver and Campbell 1989, p. 16).

With attention now fixed on diarrhoeal illness, one direction of the studies has been to narrow them down on to the pathology and physiology of the damage done to growth by repeated bouts of infection. It was apparent that infants grew well for the first three months of life. The introduction of foods in addition to breast-milk exposed the infant to increased possibilities of infection by protozoan, bacterial, and viral parasites. Complementary feeding might begin from the third month and this timing corresponded to the onset of evident growth faltering (i.e. from the fourth month). The mechanisms by which diarrhoea produces malnutrition are complex and include both anorexia and malabsorption. An early step was, of course, to identify the pathogens and the infection loads. Seasonal food shortage and high infection loads in the wet season were part of the explanation. The research moved on from study of the pathogenic organisms to the effects of recurrent and chronic diarrhoea on small bowel function. In the longer term, there was damage to the intestinal mucosa, so increasing its permeability, leading to further loss of dietary nutrients or failure to absorb them. Research continues on the interactions between malabsorption and chronic bowel infection: studies of immunological reactions to infection, gut permeability and a protein-losing enteropathy, the development of lactose intolerance, overgrowth of bacteria in the small bowel, the nutritional management of diarrhoea, a possible anabolic 'window' for treatment in the acute phase (Weaver and Campbell 1989, pp. 13–18).

A chain of cause and effect

Instead of focusing on malabsorption and intestinal mucosal changes, we can go back to the findings on seasonal variation and pursue some questions about their environment and the social behaviours which affect patterns of illness: first, the association of illness with the wet season and the introduction of food other than breast-milk. Breast-feeding protects the infant because of the much smaller risk of infection compared with giving food from a bottle or a bowl, and because there are immunoproteins in breast-milk which have some bacteriostatic properties. The pathogens found in the cases of infantile diarrhoea must have come from somewhere and the obvious immediate source to look at was any food used in complementary feeding.

Work demands on the mother, breast-feeding, and child-care

The hours worked and the distance of the rice fields from the villages, which can be up to 5 miles away, make it difficult for mothers to give as much time as they need to caring for their children. One study (Roberts *et al.* 1982) found that lactating women were active in July for a mean of 92.5 per cent of a working day 15 hours long. The younger the infant the more likely it is that the mother will take it with her when she goes out to work in the fields. Up to the age of six months babies are almost always taken to the fields; but then, and increasingly after the age of 1 year, they may be left in the village in the care of 'nursemaids', typically 4–14-year-old girls with an average age of 9 (see Fig. 2.2) (Lawrence *et al.* 1985).

Babies are breast-fed for two years. Measurements have shown that their intake of milk decreases during the wet season. At first it was thought that this was because the mother's production of milk diminished, limited by her lack of food and the exhaustion of body fat supplies in her undernourished state. But careful studies indicate that the smaller intake is almost certainly the result of infections which suppress the infant's appetite and hence its demand for milk, rather than failure of the woman's milk supply. (Prentice and Prentice 1988). The mother may also be rushed and have less time available for breast-feeding; or be too far away to come back. If she leaves her baby or young children in the care of a nursemaid, she may leave some millet gruel for them in a pot so the nursemaid has something to feed them. The longer she is away the greater the chances of bacterial overgrowth in the gruel, and the greater the risks of diarrhoeal illness. The thin gruel is not as nutritious or satisfying; the hungry infant cries for more.

Fig. 2.2   Infants more than a year old are usually left in the care of young girls in the village.

The complementary feeding may not only carry a risk of infection (perhaps especially so in the hands of 9-year-old nursemaids), it may also depress the baby's appetite through infection or by making it feel replete. The increased use of weaning foods may reduce the stimulus to lactation given by suckling.

Weaning foods

From the age of four months onwards, diarrhoeal illness became more frequent. The usual food first introduced, a millet gruel (*sanyo mono*), was prepared in the following way. The millet grain is winnowed, dampened with a little unboiled water, and pounded using a pestle and mortar. It is washed to remove the chaff, re-pounded, and sieved. The resultant moist flour is then mixed with a little cold water to form a thin paste which is added to warm water, heated until it boils, and simmered for 10 minutes.

TABLE 2.1. Percentage of food samples with unacceptable levels of pathogens

Storage time (h)	Wet season (%)	Dry season (%)
0–1	34.9	6.8
1–2	52.6	30.8
4–6	57.8	46.3
8	96.2	70.7

Source: Barrell and Rowland (1979).

As a weaning food it is grossly inadequate nutritionally. Millet gruel has less than half the energy content of milk—how much less depends on its consistency and water content. Quite apart from its defects as a supply of needed nutrients, the infant would have to take in twice as much in volume to get the same amount of energy; a stomach full of gruel might depress its readiness to suckle. And its use entails additional dangers for the baby. The mothers do not have enough millet or fuel to cook frequent meals for small children, instead they make a larger quantity and put it aside to keep. The food is cooked at 7 or 8 o'clock in the morning, and left standing in a pot for up to eight hours before it is all eaten (Barrell and Rowland 1979). In the hot climate, it makes a good medium for bacteria to grow. Barrell and Rowland found that a high proportion of food samples grew unacceptable levels of pathogens (*Escherichia coli*, *Clostridium welchii*, *Salmonella*, etc.), and that these were worse in the wet season (see Table 2.1). In other words, the earliest foods to which the infant might be exposed showed high levels of bacterial contamination with faecal 'marker' organisms and known gut pathogens.

Water contamination

Water was a likely source of the pathogens. At the time of these studies, the water they used for cooking, drinking, and washing, came from a few deep wells dug within the villages. It was heavily contaminated with coliforms. Most of the wells were not lined or fenced off. People washed beside them. The openings at the top were easily mired and became muddy, especially with the onset of the rains and the rush of water. In the wet season they could not prevent a massive increase of contamination from both animal and human sources. Although some people may have latrine and toilet areas in their compounds, or use the periphery of the village or the fields for relieving themselves, small children do not bother to keep to those areas, and there are always sheep, chickens, goats, dogs, and donkeys, and their droppings, to be seen in compounds and paths in the village. Vultures frequent the villages, sitting in big trees, or perched on a tin roof or the ground, or circling above. There is plenty to encourage scavengers. Flies buzz round mess or refuse. Ways of dealing with excrement and rubbish play their part in infection, food, and water contamination, as do people's ideas about pollution, hygiene, and washing, the values set on sharing food from the same bowl, use of the right hand for eating and the left for the toilet, methods of food storage, and so on—these are part of a complex picture in which water is crucial and linked with seasonal changes. The survey villages now do have safer water supplies, either from taps or from newly built covered wells. That improves one factor among the contributory causes, one especially important and linked with the frequency of illness and seasonal variation, but still only one among many.

DISEASE IN A WIDER PERSPECTIVE

The choice of facts about the Gambian village setting was guided by a biomedical view of the factors that are relevant to disease: more or less, we take their relevance for granted. We noted the social data on feeding, water pollution, and women's work routines because they could affect the chances of an infant or small child being exposed to bacteria or other parasites which cause infantile diarrhoea.

However, the scattering of rice fields, rules about inheritance, and Muslim piety, are not obviously or directly relevant to children's illness. In medicine it is usual to investigate disease as an individual matter, concentrating on the person and the body rather than the social setting. The questions are raised because of illness; the questions are not asked before something wrong about someone has been noticed and prompted the questions. Clinical signs and symptoms reveal the *presence* of disease, investigations of physiological changes and pathology show the *effects* of disease; they are obviously relevant. Such attention to the body stems from concern with disease and its effects. But we may need to direct attention away from the body to a wider range of external facts if the *causes* of disease, or the *prevention* of disease, become the subject of investigation. Causes of illness and questions of prevention require one to consider health, its setting, and its conditions and constraints. We must include social conditions and other people's behaviour; but how far should one look? An obvious point is illustrated by the Gambian example—that one can go on finding more and more social facts that contribute directly or indirectly to the risk of illness, but some of them do so very indirectly or to a trivial extent. The practical problem is one of pertinence and identification—how to find which are the factors associated with risk or harm and why, and how significantly. In the event of disease, there are signs, effects, direct consequences of illness to study; by contrast, concern with prevention and causes requires that we study people and situations before illness has occurred, and consider illness from both prospective and retrospective standpoints. The field of relevance opens out. Studies in the history of disease have shown in a number of fields how social conditions and human behaviour affect patterns of disease.

The scope of any investigation is bound to reflect ideas about what is relevant. If we think of disease primarily in terms of infectious diseases and genetically determined resistance, we are tempted to neglect other factors, such as environmental and social ones, which can influence illness. Infectious diseases comprise only one particular subdivision of the whole field of disease. But infectious diseases provide clear models of disease and have encouraged people to think in terms of specific single causes—a specific microbe causes a particular kind of disease—as if that was all you needed to know about causation. This simplicity is misleading. Aetiology is complex. Causal links are not always strong or simple, even in the case of infectious diseases. Exposure to tuberculosis germs does not necessarily cause illness. Some kinds of pathogen produce different illnesses depending on the age and the state of the person infected (e.g. the virus rubella, which has quite different effects on the fetus, the child, and the adult). The evidence of interactions between host, pathogen, environment, and mode of life complicates the simple view of the germ as being the cause; instead we see many factors brought together in explanation. A complete account of causation in some cases needs to take into account social environment, culture, and history, as well as biology and medicine.

A historical dimension

McNeill (1979) published an overview of the role of infectious disease in history; it provides many examples of political events and social behaviours which have affected the distribution of diseases. As Mascie-Taylor explains in Chapter 1, McNeill was concerned with populations, the major disease pools of the world, and the history of their rise and mixture. He took the history of contacts between people through war, conquest, trade and pilgrimage, and the relations between town and countryside. The ability to withstand new diseases played a significant part in conquest. Diseases could act as barriers to expansion and colonial advance. The meeting of cultures led to involuntary exchanges of disease. In the complex of factors sustaining Europe's expansion, McNeill argues that infectious disease was very significant. By the sixteenth and seventeenth centuries, populations began to grow: 'Europe's expansion is such a central fact of modern history that we are likely to take it almost for granted and fail to recognize the quite exceptional ecological circumstances that provided sufficient numbers of exportable (and often expendable) human beings needed to undertake such multifarious, risky, and demographically costly ventures' (McNeill 1979, p. 210). But major questions about the history of disease remain undecided because facts are missing or impossible to recover.

McKeown attempted to answer these questions in his study, *The modern rise of population* (1976). It was an analysis of the decline of mortality which accompanied industrialization in England and Wales. He argued that conjectures are unsatisfactory unless we can find some appropriate evidence to set before them. From 1838 onwards, rules of registration in England and Wales provided national records from which population size, birth-rate, death-rate, and cause of death could be worked out. These records provided the kind of evidence he wanted. He set out to identify what needed to be explained. He found that the growth in population came from a declining death-rate rather than a rising birth-rate. The decline of mortality was due chiefly to reduction in deaths from infectious disease. To interpret this, he identified when, to what extent, and which specific infections declined. He found that the kinds differed, each disease having its own characteristics. Airborne diseases give possibilities of exposure to risk, spread, communicability, and chances of control, which are different from diseases that are water- or food-borne. McKeown prepared graphs showing the decline in death-rate from each kind of disease with an arrow drawn to mark the date when a specific therapy for the disease was introduced. They show in most cases that most of the decline in mortality preceded the introduction of effective specific therapy. He concluded that growth in population could not be attributed to fortuitous changes in relationships between disease organism and host. The fall in mortality was not much influenced by immunization or by medical treatment before 1935 when sulphonamides became available. Medical measures were effective earlier than 1935 in the management of smallpox, syphilis, tetanus, diphtheria, diarrhoeal disease, and some surgical conditions, but they made only a small contribution to the overall decline in the death-rate from 1838. His main finding was therefore the negative conclusion that medical science had not contributed much to the decline of mortality. His main positive finding was the explanation, emerging more or less by default, that improvements in nutrition explained the rise. Mortality declined and the population began to grow before advances in hygiene were in place.

McKeown's analysis has been immensely influential. It was conventional wisdom to suppose that progress in medicine and treatment must be the chief reason for a growing freedom from disease in modern times. The shock and stimulus of his work provoked careful re-examination of his findings. His views on the lack of evidence for rising fertility have been challenged using other data (Wrigley and Schofield 1981). Szreter (1988), noting that McKeown's conclusions about improvements in nutrition emerged by default, studied the social and political developments which led to better working conditions, housing, education, health services, and stricter regulation of food quality. He concluded that social intervention had played a greater part in the decline of mortality than McKeown allowed: 'the decline in mortality, which began to be noticeable in the national aggregate statistics in the 1870s, was due more to the eventual successes of the politically and ideologically negotiated movement for public health than to any other positively identifiable factor' (Szreter 1988, p. 26). What Szreter stressed is the need to examine the *agencies* active in producing change in health rather than just the *processes* of change. He was concerned with political ideas and the individuals who motivated change. Political forces affected mortality through demands for the introduction of changes—sewage disposal, for example, better working conditions, extension of the vote, etc.

What is notable in this is the way that the various explanations for the rise in population give credit for change to different agents and forces. They focus attention on different components in complex situations. McKeown's analysis did not leave much for scientific medicine and the medical profession to boast of. The shifts of focus in analysis and conclusion prepare us for a point made by Kunitz (1987): explanations of changes in mortality and morbidity have ideological implications. The explanations can serve to justify change in action or policy. Some of them point to the role of the environment, others to individual behaviour, others to medical care. With such complex webs of cause, different strands can be picked up. Ideology or unconscious bias may influence which we prefer to emphasize. If weight is given to the role of individual behaviour in explaining patterns of morbidity or mortality, then responsibility for health may be seen as a matter for the individual ('Your illness is your own fault, you chose to smoke', 'Why should the state pay? You took the risks, you never insured yourself'). But if the environment or society is seen as the source of risk or harm (e.g. from hunger, drought, stress, pollution, overcrowding), then the individual becomes the 'victim' and the environment or society the author of the 'inequalities of health'. The locus of control is removed from the individual. The responsibilities for action shift. Kunitz (1987) refers to this contrast as one between voluntarist and determinist views of health. The contrast is roughly between active and passive: the voluntarist would argue that the individual's own choices and behaviour are important causal factors in his or her state of health or illness, the determinist that an individual's health or illness is more the result of the various external social and physical environmental forces impinging on him or her. The voluntarists may blind themselves to the impotence of the poor by assuming they could change if only they would. The determinist sees the problems in terms of access to care, poor living conditions, unsafe working places; those who fall ill are passive victims. Kunitz's point is that an explanatory theory may be more or less consistent with some particular political ideology and may contain deeply held assumptions about the nature of society, the existence of free will, the requirements of justice. Explanations and perceptions of what is relevant for understanding disease are not just matters of theoretical interest or bias: they contain implications which may affect decisions people make about treatment or the priorities in providing care.

Recognition and response to illness

In most of this discussion so far, judgement of the possible relevance of facts to disease has been set implicitly by biomedical criteria. We have not considered the views of the local Gambian people about the illnesses they suffer, or their significance to them. It has hardly mattered so far, for example, whether or not the local Gambians shared any of the scientists' ideas about bacteria, the nutritional value of gruel, or the causes of diarrhoea. The investigation could continue without asking them. Their ideas about illness were not relevant to the investigations.

However, in at least one respect their views were critical: the study depended on finding people who were ill and following the whole population in the villages over a long time. It depended on people reporting their illnesses and on how consistently they did so. What they did reflected their understanding of illness, their recognition of signs and symptoms, and their views on whom to seek advice from. Measures of prevalence will be deceptive and defective if people do not come because of fear or shame; or if they consider someone else should treat it; or if they do not see the symptoms as a medical problem. Their ideas are obviously relevant to their behaviour and to making it easy or difficult to find cases of illness. In other words, we do need to know something about their ideas about illness, and how these influence their behaviour, in order to gauge whether the data on prevalence are likely to be reliable estimates of all the illnesses which occurred during the study. An example which shows why this may be important is given by statements on the prevalence of mental illness in African populations. Estimates of mental illness among Africans, especially depression, have varied wildly; views that it was very common, or that it was very rare, have been asserted almost equally strongly, sometimes with scarcely any regard to how people with mental illness might be recognized, where they might be found and whether they would come forward to be seen by a European doctor (Field 1960; Kennedy 1973; Littlewood and Lipsedge 1982).

People's interpretations of changes they have noticed in themselves affect what they then do. To explain someone's actions, you often need to know what he thinks about the situation. Someone who feels odd (hot then cold, sweating, weak, shaky, headache), and supposes it indicates a bout of malaria, might want to act in one way; someone else, who thinks it indicates an attack by sorcery, might want to act quite differently. Odd behaviour, or marked changes of mood might seem to indicate spirit possession or witchcraft rather than illness. People's opinions on how to classify their state, the situation and its significance, the alternatives open to them, may differ. Is it a medical problem? a social problem? an attack? a crime? Even though the symptoms are the same, the factors they will consider relevant can be quite different. Knowing what the medical signs were will not necessarily explain why one went to a diviner, another to a shrine, or a third to the clinic.

Expected behaviour and the sick role

The villagers in the Gambian study see infantile diarrhoea as a medical problem and bring their small children to the clinic. By now they have long experience and familiarity with the clinic, its layout, and procedure. They know the health workers, dispensers, the nurses, and doctors; what to expect and what is expected. They must bring their clinic card and number. There is a good deal of mutual adaptation and learning behind the consultation. Any child under 3 years old brought to the clinic must be seen by the paediatrician (there is a system of screening and referral now in place at the 'gate clinic' where anyone from the study villages can come; transport is provided).

When the mother sees the paediatrician, she sits with her child beside the doctor next to a translator and describes what has happened. The translator is a highly respected community elder. He knows from long acquaintance what the medical staff want to know and also, as a wise old man of the community, he understands the implications of Mandinka thought, its expressions, and gestures. A complex process of translation and cultural mediation goes on during these exchanges: the word 'interpreter' is really very apt for his role. The translated account of the child's illness, partly the mother's statement, partly elicited and elaborated by question, is like any good clinical history. It resembles in form the sort of 'history' that would be taken in a paediatric clinic in the United Kingdom. However, the role of the translator is particularly important because the man I saw at work knew so well what the doctor wanted and could interpret so accurately what the mother meant. Sometimes he would comment, after translating her words, that the words did not mean what they seemed to say, and then explain what it was she really meant. He was acting as the cultural mediator—both for the doctor and for the mother. What impressed me, observing the Gambian mothers in this setting, was their confidence and the clarity of their answers: the relevant signs, how the child behaved, how they could tell where it felt pain, what they had done, whether they had given oral rehydration fluid as they had been taught to (supplies of the fluid are available in the village at any time of day or night; the mother only has to come to fetch it).

It would be difficult now to disentangle with any sharp precision the village women's own beliefs and convictions from the prompting, what they have learnt to say in the clinic context. The way they seem to understand childhood diarrhoea is the result of long medical contact with the clinic, the Dunn staff, and the field assistants. They have got used to an extraordinary level of care and interest in their health. But some of the mothers' under-standing is probably belief on the surface (Beckerleg, personal communica-tion). People speak about illness differently according to the context—in the clinic one set of observations is appropriate, at home perhaps another. The mother knows what she is meant to do in clinic terms with a sick child who shows signs of diarrhoea. She may also have her own ideas and choose to consult a *marabout* (a village-based Muslim diviner and healer) about the same illness.

The social construction of illness

Behaviour in illness is strongly influenced by social expectations and ideas about illness. This is implied by the phrase 'the social construction of illness' (Freidson 1972, pp. 203–331). It suggests that illness can be made and experienced differently in different societies, that we should distinguish between the biological view of disease and the way in which particular people may actually think about illness and respond to it. The contrast between biological views of disease and social constructions of illness has become the focus of much discussion in medical anthropology (Young 1982; Lock and Gordon 1988). The views of biomedical scientists and professional physicians are those of specially trained professionals. In any society, most people's views depend on various kinds of social learning and persuasion which create the cultural representations people use to cope with disease. These influence what they look for, the interpretations and mean-ings they give it, what they think they should do about it. Cultural repre-sentations can mould and modify experience in illness; they can affect outcome and treatment. The Gambian example above illustrated views which were changing. Over the years, the Gambian villagers at Keneba have become familiar with a very different social construction of illness from that encountered by Mandinka villagers living elsewhere. They have learned to adopt other patterns of behaviour in the sick role. The changes derived from biomedicine, but their construction is an understanding seen from a perspective very different from that of the medical research scientists.

Society and the mutability of illness are picked out by the idea of the social construction of illness. It confronts the biological view that nature is constant or regular in the production of disease. A phrase sometimes used in medical literature is 'the natural history of a disease'. It suggests that diseases are natural entities, almost like plant or animal species. That view goes back at least to the seventeenth century when the English physician Thomas Sydenham, a great observer and classifier of disease, likened types of disease to seeds that grow: they show a regular pattern of development and progress in those who suffer from them. In the preface to his works, Sydenham wrote (1742, p. iv): 'All diseases then ought to be reduc'd to certain and determinate kinds, with the same exactness as we see it done by botanic writers in their treatises of plants'. There is now a very long list of names for different kinds of disease. A medical textbook will contain descriptions of the typical features of each disease it discusses; the features represent the selected attributes by which the disease has been character-ized. Even though we may like to think of diseases as distinctly named entities, the disease is not an independent entity—a thing—but a selection of attributes characteristically shown by people who fall ill in this way. These can be investigated much as the characteristics of a natural species or the life cycle of an organism might be. That medical view stands in sharp contrast to the idea, implicit in the 'social construction of illness', that our experience and understandings of illness are largely created for us by cultural representations, learning, and persuasion.

The representations are persuasive. In any society people are bound to depend on others, not firsthand experience, for most of what they know about illness. Medical knowledge in the sense of the ideas local people have about illness, what they believe about its causes and treatment, the impor-tance and significance they ascribe to different features of illness, is there-fore highly contingent on the society and culture in which they have been brought up. The moulding of illness by society or culture can have both cognitive and behavioural aspects. There are examples of strikingly diver-gent views about what is or is not illness: measles regarded not as illness but as a stage of development (Hong Kong: Topley 1970); nocturnal emission of semen seen as an illness (*dhatu* loss: Obeyesekere 1976); continual differences of opinion about mental illness, whether shamans are schizo-phrenic or sane (Ackerknecht 1971); the status of depression as a disease (Kleinman and Good 1985); culture-bound syndromes with their particular histories bound to certain times or places—*latah, imu, koro, amok, witigo, susto,* nostalgia, acedia, anorexia, shell shock, neurasthenia (Simons and Hughes 1985). However, far more significant as evidence of the cultural moulding of medical opinions than these examples of occasional highly divergent or odd views, I think, is the ordinary endless diversity of opinion about kinds of illness, their significance, and which are bad signs or symptoms, and what ought to be done.

Freidson (1972, pp. 205–7) uses the concept of the social construction of illness to emphasize the distinction of physical from social reality, a distinction of the physical facts from social representations of those facts. He turns the idea back on to medicine itself: 'In the sense that medicine has the authority to label one person's complaint an illness and another's complaint not, medicine may be said to be engaged in *the creation of illness as a social state which a human being may assume*'. In our society the medical profession is officially recognized as the authority for what is or is not illness: by virtue of being the authority on what illness 'really' is, medicine creates the social possibilities for acting sick. A layman may have his own 'unscientific' view of illness diverging from that of medicine; members of another culture may have views of illness that differ from those of Western biomedicine. Kleinman (1980) similarly puts the concept of 'the cultural construction of clinical reality' at the centre of his comparisons of illness and healing. 'The behavior of the ill varies from one culture to another, very often independently of disease, and constitutes a reality in itself. So does the behavior of the healer vary from culture to culture' (Freidson 1972, p. 207). In effect, when it comes to perceptions of the situation, views and theoretical standpoints, these authors say there is not one reality but many. This is certainly so in a descriptive sense: people's perceptions, their experiences, and their reactions differ although we may choose to identify the disease (in terms of biological facts about the illness) to which they react as the same. One says: 'This is an attack of malaria', the other 'This is an attack of sorcery'; the biological facts—fever, shivering, and sweating—are the same. European medical views on what does or does not come within the field of disease have a history and have changed. We are dealing here with social reality. The responses might be very different depending on what the people involved thought.

Behaviour in illness: the Gnau

The behaviour of Gnau-speaking people when they are ill provides a contrast to that of the Gambian mothers and patients who actively seek treatment. The Gnau-speaking people live in three villages in the West Sepik Province of Papua New Guinea. When I first stayed among them (1968–70), they often relied on non-verbal show more than speech to convey that they felt ill. This involved social withdrawal, refusal of food and conversation, self-neglect and inertia, and dirt and dust left on the skin. Such behaviour was regarded by them as appropriate if they had decided that their symptoms might be those of a serious illness (Lewis 1975, pp. 95–153). The behaviour could be misleading as a guide to the nature of their symptoms.

Some effects of disease may be universal, but most of the effects on behaviour are not precisely determined by the specific disease. Although the body may break down in similar ways whatever the language or culture of the person affected, most diseases are imperfectly and not universally correlated with specific behaviours. The scope for a learned or controlled response varies, of course, with the kind of disease—compare, for example, a broken leg, an epileptic fit, painless swelling of the ankles, a bout of 'flu. Some effects may depend just on what the disease does to the body, but most are likely to be partly related to that and partly to social training and expectations. Mechanic (1962, 1968) showed how the perceptual salience of signs and symptoms depends on more than just their biomedical characteristics: apart from the severity, pain, or frequency of symptoms, the extent to which they were obvious or disabling in work and social activities, the availability of information, the distance of treatment resources, social stigma, competing needs, the psychological and material costs of seeking treatment—these were all factors which might affect what someone did. Mechanic's detailed analyses of the social and psychological factors which pushed people to seek help, or deterred them from doing so, were influential. His work was primarily concerned with North American urban contexts. Fabrega (1974) suggested that cross-cultural comparisons of behaviour in illness would help to discriminate the universal from the culturally specific in responses to illness; and show how big a part social and cultural factors play in modifying behavioural responses to particular diseases. Medical science provides a relatively refined, precise framework for describing biological systems, their structure and function, and how they break down: something equivalent should be devised for the social and behavioural responses. Fabrega and Zucker (1977) advocated careful description of behaviour in illness as the means to construct a 'grammar of behaviour in illness' (it would involve recording gestures, posture and pain-related immobility, disinclination to eat, stopping work, time spent in bed, actions directed to obtaining advice, remedial actions, etc.). It could provide a framework for recording the sequence of changes during an illness, and for finding what is common and what rare in responses, and for comparing the patterns and peculiarities of different cultures.

Clinical medical observation is usually focused rather narrowly on the sick person. However, it would be a mistake for anyone trying to understand a sick person's behaviour to consider only the individual in isolation. People modify their behaviour according to who is present and what they want to let others see. Behaviour in private may often differ from behaviour in public. There is a voluntary element in hiding or revealing illness. The response may be chosen to alert or inform others because the sufferer needs help. Or, on the contrary, someone may want to hide signs which are shameful or stigmatizing. The ways to convey pain or to gain sympathy are subject to cultural conventions. Ideas of appropriate expression differ: calls for help, cries in pain, stoic understatement, and so on (Zborowski 1958).

The Gnau relied on non-verbal show. Sometimes the behaviour looked ostentatious. Men might act being ill too conspicuously, usually this was because of a suspicion or fear about the implications of their symptoms. In any case, Gnau men tended to pay attention to trivial illness more readily than women. It is easier for them to take time off to stay home ill. They are not bound in the same way as women by daily demands to feed and cook for the family. If a woman showed her illness very conspicuously, it was more often done to make clear just how ill she felt, it seemed more like an appeal to be let off domestic chores—her behaviour was less likely than a man's to be linked to a special fear or explanation (Lewis 1975, pp. 239–44).

The Gnau men and women and the Gambian mothers illustrate in different ways the general point, that behaviour in illness is the outcome of a complex mixture of forces: biological effects of the illness, local explanations, cultural conventions about what to do, responses to the other people present in a particular social field. Fabrega's focus on behaviour in illness has been eclipsed by the attention given to cognitive rather than behavioural aspects of illness—a focus on its cultural meaning and interpretations of its significance. However, behaviour is the common ground bringing together biological, cultural, and social components of illness.

DIAGNOSIS AND THE CLASSIFICATION OF ILLNESS

In many societies, people classify illness along different lines from those used in Western biomedicine (WBM). In the WBM discipline, signs and symptoms are indicators of disease; the disease can be inferred from them (plus special tests). Although Gnau people could describe pains and symptoms, they did not usually volunteer much detail about them. Sometimes their answers to direct questions were vague or unreliable. Descriptions of symptoms, such as I collected among the Gnau, do not automatically lead to a classification. Classification implies some organization of knowledge. The criteria and features used in it are chosen for some purpose. Study of the principles behind the classification should reveal what the people who use its categories consider to be significant in that field. The criteria they use have to be found out. Relatively few diseases have striking 'pathognomonic signs'. People do not always classify illness primarily by body signs and symptoms. And in many individual situations, people are less concerned about applying a precise diagnostic label than interpreting its personal significance.

A conceptual distinction between disease and illness

Freidson stressed the distinction of physical from social reality (sceptics might go beyond his distinction to question the reality and status of the biological 'facts'). For the time being, however, we can keep in mind a conceptual distinction between 'disease' and 'illness': the distinction uses 'disease' for the biomedical view, kinds of disorder distinguished on the basis of biological facts, with physiological, psychological, and anatomical criteria of abnormality which should, ideally, be recognizable objectively; and 'illness' for those corresponding changes of body or mind that people choose to identify as undesired and abnormal, the changes so recognized may vary according to culture and with individual; 'illness' refers to the experience and meaning of perceived disease. The Western biomedical views on the classification of disease are often invoked as the standard for universal use: with varying stress, it has been implied that they are objectively framed, free from cultural bias, scientifically based, and correct. But it is also clear that WBM classifications change, current knowledge is incomplete, normal standards for some characteristics have to be revised, certain ideas have been shown wrong in the past and no doubt others will be in the future, scientific ideals are not always evident in practice. The distinction of 'disease' from 'illness' has therefore prompted questions about the special status given WBM, and raised issues of objective and subjective criteria, scientific fact and cultural relativity, universal and local cultural standards, 'etic' and 'emic' categories, notions of normality, ideal and average standards of normality, social or cultural standards versus individual criteria. Exact definitions of the distinction have varied (Fabrega 1974; Lewis 1975, pp. 146, 355; Eisenberg 1977; Kleinman 1980, p. 72; Young 1982), but one way or another, the contrast between 'disease' and 'illness' has been very widely used.

Description and interpretation

Diagnosis begins with recognition of some change and identification of what sort of change it may be. It is the first step towards understanding it and may lead on to an interpretation of its significance. The first stage involves noticing the change and identifying it in terms of some classification which will usually imply something about its significance. It may be a matter of identification through description or it may go beyond simple description to include interpretation. Classifications of illness often show complex mixtures of description and interpretation. 'Fever', for example, is a descriptive category which could include someone with 'malaria' or 'measles'. Malaria and measles have clinical characteristics that can be described; the categories also imply interpretations of their respective specific causes. 'Diabetes mellitus' is a diagnosis which, although fairly precise, conveys little about the particular signs or symptoms of someone affected by it; the term is a diagnostic conclusion, interpretative rather than descriptive. Recognition may convey immediate implications of significance. Different categories and classifications may tend to focus attention on descriptive features of the illness or on its interpretation. If interpretation, it may suggest things about the underlying mechanisms of illness or about the origins or cause of the illness. The first step, therefore, of classifying the problem can be crucial for orientating attention and decisions about what is relevant for understanding it.

Someone who suspects that his fever comes from sorcery will act differently from someone who supposes it is malaria. He has classified the event differently; the facts that might be relevant change. I have stressed that we learn what to think and, in general, we tend to accept the prevailing ideas in our society as our own beliefs. Most of our knowledge is learnt from others, not firsthand experience. This is the case with illness and we should expect people's responses to reflect their cultural background. People's responses grow from learning and depend on recognition, opportunity, and experience; recognitions of pattern, features, and similarity, are at the basis of diagnosis and classification.

Standards of normality

First, it is necessary to recognize the change, a difference from normal. Recognition depends on attention and standards of normality: these may be set both individually and culturally—individually because each of us knows how we usually feel and what demands we can normally expect to make on ourselves and our bodies, culturally because social life sets some of these demands and puts priority on different sorts of performances, establishes criteria of approval, and normality.

The question of normality is difficult. In medicine, the criteria are both statistical and ideal. The functioning of different organs and systems is conceived in relation to ideas of adequate or proper functioning for the maintenance and integrity of the whole organism. Transmission of genes, through survival to maturity and reproduction, constitutes the final aim or purpose in the teleology of biological reasoning. But some functions vary with age and sex, and so the standards to apply must vary. If we take the average as a guide to normality, it has the advantage of being measurable, it avoids exposing our selective prejudices about what is desirable and what is not. But a purely statistical view of abnormality ends up identifying the abnormal with the unusual. In the United Kingdom, having good teeth is unusual. Great strength, great skill at chess, etc., are unusual. So we must import some ideas of what is relevant and desirable for health; hence, the ideal aspect. However, ideals involve what ought to be and they resist exact definition. They cannot be identified as straightforwardly as an average can. Absence of disease, and adequacy for health, prove less ambitious criteria for health than ideal health conceived in terms of perfection. The constitution of the World Health Organization aspires higher: 'Health is a state of complete physical, mental, and social well-being and not merely the absence of disease or infirmity'.

The theory of evolution might seem to link abnormality with malfunction through natural selection; to imply that the conditions we call normal in the sense of average will be functional, and that most gross departures from them will be malfunctional. But serious difficulties arise when we consider human populations in which social conditions and conventions may strongly affect both the adaptiveness and the standards of what is normal or average. We face great difficulty in specifying what are 'ordinary circumstances' because of the degree to which people reshape their own environments. 'Ordinary circumstances' differ greatly by country and they can be changed. People modify an environment to suit their wishes as well as their needs. Whether some attribute brings advantage or disadvantage may change with place and time. Findings become average which are not ideal—tooth decay is an example. People may learn how to protect themselves so that those traits which would otherwise put them at a biological disadvantage survive. Many characteristics are not overwhelmingly or strikingly malfunctional, but are only relatively so, requiring various other stresses to produce evident sickness. Or they are graded characteristics, like blood pressure or serum cholesterol levels, which fall within an average range on a continuum. They begin to be statistically associated with manifest sickness toward the upper or lower edges of the range: the associations with sickness are not inevitable but depend also on other contributory or predisposing factors. The identification of standards is then very difficult. The harmful effects of some malfunction may be more or less severe, depending perhaps on the environmental or social circumstances of the person affected. The effects are, in other words, relative to circumstances. Normal functions obviously are relative to the species. But normal function can also be relative to individuals and to circumstances.

Talcott Parsons (1951, chapter 10) presented a view of medicine as being a social institution: he wrote that illness could be looked at sociologically as a form of deviance potentially disruptive to the functioning of society, and medicine could be seen as the means to cope with and control it. He suggested that we could define health from a social standpoint as the state of optimum capacity of an individual for the effective performance of the roles and tasks for which he has been socialized. If illness is defined on the basis of standards of adequate functioning needed for the performance of activities expected from members of a society, a single uniform health criterion would have to be replaced by norms having different meanings for different people. The implications are that social requirements may set medical standards. In the past, medical authorities have perhaps misrepresented their own social attitudes or prejudices as the standards for normal biological function. Or their standards were in fact set by political demands, but stamped with authority as medical decisions (the classification of political dissidents as being mentally ill is an example). However, social deviance or maladjustment is not identical to disease. Conformity to social pressures and rules is not necessarily a sign of physical or mental health. If deviance and disease are identified, then disease becomes a tautology for social maladjustment, and health a tautology for social adaptation. If disapproval becomes the criterion for disease, then the diagnoses of disease will depend on which group expresses an opinion. Certainly, social conditions may mask or mould the signs of sickness, make a handicap more or less of a disadvantage. But if the ability to hold a job is taken as a sign of social competence, and therefore of health, judgements about the health or sickness of individuals can depend on the state and saturation of the employment market.

People tend to want to take their own behaviour as the standard for normality. They may want to call those who differ abnormal; just as they find strange customs sometimes unhealthy or repugnant. Education and the experience of a particular way of life will set local people's standards and expectations. The meaning-centred approach to illness allows us to grasp the actor's point of view, but it need not preclude others from assessing the validity of the standards which someone uses to judge the signs or symptoms of disease.

The concept of accident and the interpretation of natural events

The following is a case history of Berau.

Berau is aged about 40, a married Gnau woman in the village. She fell heavily on to her left wrist when a big load she was carrying up a steep path caught on a branch. She was carrying the load with a headstrap and her head was twisted as she fell. Her wrist was badly swollen but with no obvious fracture. She made no fuss about the pain. I tried to splint and strap her wrist. She soon took it off and asked for liniment. Three weeks later her wrist was still swollen and painful and I began to worry that I had missed a Colles fracture. She said the wrist still hurt, especially when she tried to chip sago or climb a tree to collect leaves! Then she disappeared for three weeks. She went to stay at a distant bush camp. The wrist got better. No deformity. She was right. It was just a bad sprain.

The injury and the pain were intelligible to Berau. The fall was an accident. People's previous experience influences them about when to worry, how they respond to the pain, how hard they try to find an explanation. The sequence of events, the path and the branch, were enough. The Gnau do not, so far as I know, have a word which exactly matches the European word, 'normal'. Clearly, considerations of commonness or triviality, of what can be expected or understood as a direct visible sequence of dependent events, lay behind statements (such as Berau's view) that something had 'just happened' *gipi'i* (literally, 'nothingly'), or happened without intention or some cause or contriving behind its occurrence—the word used is *diyi* ('without cause or purpose').

European concepts of nature and natural events are peculiar and complex (Collingwood 1945; Lewis 1975, pp. 196–202; Williams 1976, pp. 184–9). We speak of illness occurring as a result of natural processes. It is not just a matter of accepting that certain illnesses just occur, or expressive of a lack of concern or interest, but part of a view consciously held and applying both to trivial and serious ailments. Of course, individuals in our society may want to explain some illness further in terms of religious or moral meanings, but the general assumption prevails. It underlies a view that illness can be investigated in a purposeful search for general rules or 'laws' to link together the observed phenomena. What was first meant by 'nature' was some essential principle, the inborn element which might, for instance, distinguish natural from supernatural or contrived happenings, or the ordinary from the magical or miraculous. Secondly, there was 'nature' viewed as the existing scheme of things: the natural phenomena surrounding us, nature around us, this world by contrast with some other, with heaven or hell, the supernatural. Thirdly, there was nature viewed as a universe acting according to rules or laws, the regularity of nature. The assumption of uniformity in nature is not necessarily formulated in the same way or shared by other societies. The general view of nature in which events must obey impersonal laws is one that produced the idea of regularity, the constancy or uniformity of nature that sustains the division between what is thought possible or impossible. It influences what kind of attention is paid to rare or extraordinary events. It differs greatly from the view that various forces may intervene irregularly, variously willing or intending events. Selective attention is strongly influenced by what is understood of possibility and impossibility. People try to make sense by looking for patterns or relationships between the things they notice. Over illnesses, it might be a search for the patterns in body signs or in the circumstances or events surrounding illness: syndromes of disease or syndromes of circumstance explaining cause.

Explanation, the scale of society, and experience of disease

If a change is noted, we may worry about what it is, what it means for the future. Is it a medical problem or something else: punishment? accident? attack? A more detailed diagnosis can be pursued. People draw on their experience and consult others. Experience may improve skills. The ability to recognize a condition grows with familiarity. Whether a pattern can be seen in the relationship between observed signs or between sign and symptom, or in some syndrome or circumstance, may depend on having enough cases to observe, or on the salience of features or of timing. Comparatively few illnesses have really distinctive signs. With most diseases it will require observations of many examples before any regularities or patterns are noticed—the 'natural history' of the disease, the way the illness develops, its progress and outcome, the typical settings in which it occurs. Classifications of kinds of illness or disease are developed from such observations. Common illnesses become familiar, but rare ones do not: their signs and symptoms fit no particular established pattern. Small numbers of people limit experience; there are few cases to see. One person's experience will differ from another's. It is often necessary to consult others. Classification is concerned with identifying similarities and differences—experience, repetition and familiarity increase the information on which to draw. So classification and nosography may be the domain of specialists, whereas the ordinary person is not much interested in details except in so far as they directly concern him- or herself.

The significance of an illness for the individual

Illness can single someone out. From an individual point of view, a diagnostic label provides only partial identification. It identifies the category, the general type, but not the personal and individual features of the case. Each illness is unique: it comes at a particular time, beset with its particular contingencies and problems. It is an episode in someone's life, part of his or her unique personal history.

The case of Silmai with a bad headache:

He was sitting crouched over a fire beside his house. Silmai is in his fifties, a wily man, also sardonic, intelligent, and effective in authority. His head was bent down on to his knees, his eyes closed, and although he heard us come, he made no movement to look at us or acknowledge our presence. Saibuten asked him if he was sick. In a gravelly voice, no change of his bent-up position, he said he was, and something about relatives on his mother's side cutting down a tree. He raised his furrowed, bleary, haggard face, and laid his cheek sideways on his knees. He had a headache and ached generally. Then he said he had dreamed of Galwun and

Sukadel, matrilaterally related to him. He had sent for Sanawut to come and spit over him because he is also matrilaterally related. There were two splotches of red betel juice from the spitting on his shoulder blades. We sat with him for a while; he, still crouched, occasionally snapping his fingers at his forehead, occasionally giving short hard blows with the heel of his hand against his forehead (that must have hurt!). He didn't talk any more. He growled at a dog sitting by him. He got up to urinate, came back, lay down to sleep. Creased brow, haggard look—a bad headache. Saibuten explained that Silmai thought they might have been cutting down a big old tree on his mother's brother's land. (The Gnau liken the relationship of a sister's son to his mother's brother's clan to that of a tree growing up from a piece of ground. This image associates the sister's son with the tree, the clan with the ground. So perhaps when a tree growing on the mother's brother's land is cut down, the sister's son will feel sympathetic pain. A mother's brother has the power, they think, to harm his sister's son by calling on his ancestors and putting a spell on such tree, then cutting it down. As the tree falls so will his sister's son be struck down.) Saibuten was alluding to these ideas in his explanation. The idea had passed through Silmai's mind because of his dream. But Silmai was better next day.

The illness can therefore be only partially pinned down by a diagnostic label, such as headache, which says nothing about the individual circumstances of the case. Classifications are bound to focus on the characteristics of categories and to ignore contingent individual features. Sometimes those contingent features are the factors that determine whether the illness is serious or not for the individual or the family concerned. Classifying the illness is different from trying to understand its personal significance or trying to explain it. It might be the way the illness impinges on work or social life. An illness may disrupt plans or hopes, its significance depending more on who or what is affected rather than on the particular kind of disease. Its personal significance and timing may dominate attention and attempts to explain why it happened. For the individual faced with illness wants to know what it may signify. Is it meaningful or meaningless? Is it just part of how he is? Is it his fault? Was it sent? Or just chance? The significance of the illness is not dependent solely on the nature of the disease, it may depend on quite different social and personal considerations.

The timing of illness may give it significance as in the following case in which the young wives of two brothers argued, and their mother-in-law fell ill.

The young wives of two brothers had a public argument. They fought with both insults and blows about a sago palm they both wanted to work on. The same evening, the wife of the younger brother ran off back to her parents where she stayed for the next week. While she was there, they said she was ill and she behaved so. Not ill or injured in a particular way, nor treated for a specific cause, but acting ill and low in mood. She looked as if she might be ill. Meanwhile the mother-in-law of both of them developed a chest infection.

When the mother-in-law fell ill, she blamed her illness on them for quarrelling, and suggested that a spirit of her husband's lineage made her ill because it was cross at the lack of harmony between the wives of its descendants.

Such ideas are likely to affect sympathy for the sick person. But why she, rather than one or both of the daughters-in-law, should have suffered was ignored. There was disharmony among the families of that part of the lineage, one member of the group had fallen ill, the timing fitted, that was enough. It was true that one of the daughters-in-law had earlier run off back to her parents; perhaps that was partly in shame, partly anger, partly fear of being struck by a lineage spirit if she stayed, partly for comfort and sympathy. At her parents' home it was as though she was there because she was ill, at least she was there to avoid the danger of illness. People say a spirit might be provoked by the sight of someone. She may have felt at risk near the place of her husband's spirits because of the fight. And there was the element of hoping to deceive a spirit into thinking she was already sick or harmed. People say that to eat or chat and laugh normally when ill would provoke the spirit; it would think you took no notice of its warning. There may be an intended deceptive element, they recognize, in the display of sick behaviour. Gnau people leave room for ambiguity and different interpretations when they explain illness; often they describe the facts and circumstances but draw no definite conclusion.

CLASSIFICATION AND EXPLANATION

Classification is based on the recognition of similarities and differences between cases or instances. Things are classified according to some principles or with certain purposes in mind. Focus on the criteria used to discriminate between categories may show more clearly what information people convey by their categories. Frake (1961), Good (1977), and Nichter (1989, pp. 85–123) provide illuminating examples of different analytic approaches to the meaning of other people's categories of illness. From field-work in Mindanao, the Philippines, Frake (1961) analysed the diagnosis of disease among the Subanun. It was a seminal demonstration of the componential method of analysis applied to illness. He set out to learn to apply their terms as they did: he was not trying to match their categories to those of Western biomedicine (WBM). He isolated the distinctive features they attended to in diagnosis; he was concerned with their criteria for inclusion in a category or exclusion from it, with the taxonomic hierarchy and ordering principles of their classification. He produced a classification of disease in their terms, showing what their categories denoted. He did not go into the connotations of the terms or individual variations in how people used them. Good (1977) argued that these issues need study. Good's essay on the meaning of the diagnosis of an 'illness of the heart' explores the broad network of meanings and connotations of this illness in an Iranian community. He studied associations of the diagnosis with causes and social circumstances and settings. He investigated the frequency of mention of different elements and by whom, how the different elements of meaning were interlinked and related to each other. Taxonomic methods like Frake's are static and formal. The classification is too sharply cut and unambiguous. It cannot show adequately the interplay between the diagnostic and other related features, or provide insight into the varied ways people understand and use the terms. Nichter (1989, pp. 85–123) reviews these issues in his study of the language of illness, its use in interactions, and the negotiation of diagnosis.

The Gnau approach to diagnosis was usually by way of discussion or conversation about the person who was ill, in private it might involve questioning, and occasionally special techniques of divination. But they did not in general assume that the cause of illness would be revealed by examining body signs and symptoms. In trying to diagnose and explain illness, they were much more concerned to find out about the circumstances and social situation of the sick person. This made their approach to illness seem very different from that of WBM. Their diagnostic categories did not usually correlate with precise clinical features as WBM would see them; it would often not be possible to guess what part of the body was affected or how the person felt just from knowing which diagnosis they had come to. The difference in the diagnostic value they attached to clinical signs helps to explain why they seemed so vague and inaccurate about their symptoms and signs. In the medical system we are familiar with (WBM), clinical examination is the basis of diagnosis. The guiding idea is that the kind of disease can be identified from the symptoms and signs and they will lead to discovery of the seat of illness and its cause. It is why the doctor first takes 'the (clinical) history' then does 'the (clinical) examination'. The method is semiological, the signs are clues to the diagnostic answer; signs on the surface sometimes serve to identify what is internal, hidden or invisible; and from a diagnosis it may be possible to predict outcome and choose treatment. The basic assumption is that nature shows regularity in the production of disease.

Possibilities of knowledge: technology and the clinical gaze

The clinical methods of diagnosis in WBM have, of course, an evolving history; they have depended on advances in knowledge and equipment. Inspection, palpation, auscultation, the use of the stethoscope, laboratory tests, and so on, became established parts of clinical procedure at different times. Their use rests on complex assumptions about the value of the information they can give but the value is by no means self-evident to someone who does not share the same ideas or premises about illness. The Gnau, for example, knew how white doctors examined someone ill but they did not like the method or see the point of it.

We may take WBM methods for granted yet be surprised to learn when it was that doctors began to examine their patients, to expose them to touch, palpation, percussion, rather than just talk to them, and observe their outward appearance and manner. After auscultation was invented, doctors gave much more attention to their ears in listening to sounds the body made. Innovations in chemistry, lens-grinding, and electricity, altered the possibilities of knowing what was going on inside the bodies of the sick. A particular discovery in some other field, for instance the thermometer or the electrometer, led, but not immediately, to particular changes or precision in identifying some features of illness, sometimes quite altering the old perspectives in which the diseases were seen. The diagnosis of heart disease changed with the advent of the electrocardiogram. And so on, through X-rays, blood chemistry, etc. (Reiser 1978). Technology thus altered the clinical gaze, and the possibilities for knowing about disease. The Gnau and people like them share very little of this. It is not surprising that the values they attach to clinical signs are so different, or that the ways they set about understanding illness go in other directions.

Of course, the Gnau were concerned to diagnose the illness and find a way to treat it. They too tried to identify its cause. But they did so more by conversation, questions, an examination of circumstance, discussion of events and social analysis. Most of the causes which they were interested in could not be detected just from bodily signs. The categories of Gnau diagnosis were not elaborated into many disease syndromes or clinical types of illness, but rather they focused on identifications of cause. An attempt to compare their diagnostic categories with those of WBM brings out how different they are. For example, one cause of serious illness that they recognized was a spirit named Malyi. The spirit has associations with certain places, foods, and activities; certain lineages and certain individuals had strong ties with it. During a long stay with the people I saw 13 cases of illness in which the diagnosis of Malyi was made: but in WBM terms the illnesses they suffered were varied and they included the following under the diagnostic label of illness caused by Malyi: acute renal failure, arthritis, congestive heart failure, pneumonia, perirectal abscess, common cold, aches in hip and thigh. The diagnosis of the spirit Malyi was an inference about the cause based on other evidence than that given by the bodily signs and symptoms. Such a system of diagnosis obviously makes it very difficult to connect their causes with WBM diagnostic categories. They are unlikely to find that regularities in the clinical features of illness correlate with their kinds of diagnostic category.

Causal explanation among the Gnau

The Gnau looked outside the body to identify the cause. The body may reveal very little of what is going on inside. People do not necessarily suppose that clues to the explanation of an illness must be found within the body rather than outside (Young 1976). Forces which lie outside the body might act to make a person ill. They might be invisible. The problem then is how to find a connection between the person who is ill and the circumstances or things which made him ill. The unpredictability and distress of illness was a common spur to the Gnau to search for an explanation in their ideas about risks in social life and the environment round them. Illness acted sometimes as a catalyst to thought, calling beliefs into question. They tried to link observations of timing, behaviour, and events.

Although occasionally they used special techniques of divination, the usual Gnau approach to explaining illness depended on linking the patient and his circumstances (which were things to be observed) with attentions from certain agents or causes of illness (which were not ordinarily visible). Their ideas about these causes included a variety of theories about spirits, sorcery and taboos, and assumptions about their powers. They often deduced the relationships between facts through analogies with their experience of motivating forces in social life. Faced with explaining a case of illness, the Gnau considered the events of illness in a matter-of-fact way first, trying to deduce what might be the likely cause. I sometimes found it hard to tell whether I had just heard a plain description of recent facts, or a statement by implication about the mystical causes of an illness. They worked out possibilities by considering the timing and circumstances surrounding the illness. Their conclusions would perhaps light on certain components of the situation as significant for understanding how and why the illness had occurred. The reasoning process is similar to that which we use in trying to decide why an accident occurred. We consider what risks attach to different components of the situation (Lewis 1975, pp. 265–6). Their diagnoses derived from syndromes of circumstance rather than syndromes of clinical symptoms and signs. But their assumptions were different.

Gnau notions about the presence of spirits, their attachments to place, their movements, their temper and interests, animated the landscape in which the people move with varied hazards. When people worked in gardens, for instance, they were cautious and observed proper behaviour because the spirits were thought to be present and watchful, indeed after certain events like planting, activated and aware. A large fern and the snake creeper, for example, might be avoided because they could have a watching spirit attached to them as a man would be to his house or sleeping-place. In cutting things down they saw themselves as cutting down things which might belong to certain spirits, or in which these spirits had an interest. The same went for crops ready for picking or harvesting, and prepared sago. When cutting was involved the imputation was that spirits might be annoyed or angered at the damage to their things; when things were taken, the imputation was of their anger at the loss. These ideas were part of ambivalent views about the power of spirits which also had its other face— that of benefit, securing crop growth, and abundance, aiding people's efforts to produce their food. The power of the spirits was essential to successful gardening activities. The benefits must be sought and so the risks must be run. They were not wholly predictable and avoidable risks (Lewis 1975, p. 314).

Much of the time, ordinary experience provided ideas and analogies for making sense of illness. This came out in the language they used, especially the verbs linking cause and patient. They would speak of a spirit which caused someone to be ill as crushing, holding, fastening round, tying up, holding tight, or pulling on the person afflicted. The verbs express the sense of a patient crushed down, constrained, restricted by illness. The spirit could strike or shoot the patient (the idea of attack, the hurt, disease the enemy), or stay in him, or go down into him (the idea of unwanted presence, disease the intruder). Sometimes it seemed they took this idea of entry literally, for instance, when they sucked out 'arrow points' shot in by the spirit, or when they tried a stinking creeper round the affected part to stink out the spirit. They would sear the patient's skin with flaming coconut fronds to drive it out by heat; or startle it with blows and bangs. Simple verbs described how a spirit would take notice of someone: it would spy him out, put its eyes on him, say his name, call out to him, smell him. Both speech and behaviour showed what they thought. These views and deductions about the causes and processes of illness reflect their attempt to find some order, to explain events, to predict them and control them.

Food was a common focus of attention in their diagnoses. It may seem obvious that food should be a subject of concern. Food enters the body. If illness may result from some harm inside the body, food is a vehicle for it to get inside. The idea of poisoning by food was there but it was not the only sense or the main sense in which food was linked to cause. What also prompted their concern were proscriptive rules about specific foods (food taboos) and also the ideas I referred to above—that spirits watch over food. The entry of the food into the victim's body could have singled the person out. Sometimes it might be food eaten by the mother which gave a reason for her baby's illness. Food eaten by other people could make someone sick feel worse when they came too close to him or her because the spirit associated with the food would notice. The spirit might attend to those who eat food it looks after and follow them and their movements and contacts. Food taboos provided an elaborate classification of dangers; some of the rules depended on kinds of food combined with features of the person's stage of physical growth, maturity, or achievements; others depended on gender or kinship relations (Lewis 1980, pp. 146–65). The rules have effects on what people eat and on food distribution, but they also make people think. The rules imply a particular understanding of the interrelationships between human beings and what they do and natural objects, creatures, and plants—a classification and an understanding of nature, its regularities, and the principles active in it. Food is not there simply to be gulped down. The valencies and values, associations with the land, the grower or the hunter, myth and spirit, call for discrimination and add to the significance of the food. Illness is often a sharp stimulus to think about things eaten and actions. Illness sometimes provokes questions of belief or makes people recall events and see actions in a new light. The rules and taboos are meant to guide and protect people, they have a preventive purpose.

The personification of spirits based on an understanding of human behaviour allows the Gnau to make guesses about motives which might actuate spirits in causing illness. But despite their personification, spirits were not considered persons as people are. They did not think they could judge or understand the motives of spirits as well as they could those of other people. At times the spirits seemed as capricious as the wind. So people did not assign responsibility and blame to spirits with the same moral indignation as they did to people who were thought to have caused someone's illness by sorcery.

Moral judgements of illness

People's attitudes to someone's illness are influenced by ideas about its cause, especially by implications of responsibility or by the injustice of suffering. Some conditions are strongly linked with particular behaviours or circumstances. Moral judgements are almost inevitable. One reason is that there are some behaviours which excite strong social disapproval. If a particular diagnosis implies or calls to mind such behaviour, the diagnostic identification may stigmatize the person affected. An obvious example would be strong moral rules governing sexual behaviour and the stigma or shame associated with sexually transmitted diseases. Another reason may be that the very unpleasantness of some diseases—those which disfigure or handicap or hurt—has often prompted people to try to make sense of the unpleasantness and suffering in moral terms, those of justice or injustice, rewards for good or evil. Explanations which link cause and condition often suggest the lines on which people would expect to attribute blame; for instance, illness as punishment. Even if the explanation is not necessarily a matter for shame, it may bring grounds for judgements about someone's foolishness or thoughtlessness (if the risks should have been considered). But judgements are often left ambiguous or unstated. In the case of the mother-in-law who had a chest infection, the explanation was left unclear. Ambiguity or complexities sometimes help to meet the difficulties of reconciling illness with justice. They increase the alternatives of explanation, allow people to attribute some significance to events which otherwise would have to be left meaningless but distressing. Too rigid a system would destroy belief.

Without alternatives or ambiguities, an inflexible link between cause and condition is likely to bring problems. If a certain condition is associated with shameful behaviour, it may be hard to credit it in every case. In many societies, illness has been partly linked with law, religion, and morals; as if people had seized on the nastiness of illness to turn it into a sanction to reinforce rules of conduct. Leprosy, venereal disease, and AIDS provide examples in Europe of diseases to which people have attached blame. Examples can be found in many societies. Nature in the form of illness was taken to uphold morality. Justice and fairness become involved. The problems arise when people try to make too much sense of a medical event or condition in moral terms, and the implications become unacceptable or seem quite wrong or unfair for the person concerned.

Labelling and stigma

It can become especially unfair when the link is strong but false, or the stereotype obscures the facts. Many WBM medical categories are closely identified with their causes, for example, malaria, and iron-deficiency anaemia. The cause may be the main association of the diagnosis. It is easy to confuse a condition and its cause. Diphtheria the illness and diphtheria the germ are scarcely distinguished in ordinary thought. The label 'diphtheria' describes a way of being ill, it also refers to the germ which causes it; the germ is a microorganism that can be grown in tiny colonies on an agar plate in the laboratory. But the germ is not the disease. Just as an idea of disease is so strongly attached to the word 'diphtheria', so attitudes to a condition, or beliefs about its cause, can become attached to a diagnostic label. The diagnosis immediately calls to mind the typical conduct or the situations which are believed to expose one to risk; and it is behaviour that would generally be condemned. Syphilis is a sexually transmitted disease: the risks of acquiring it are increased by frequent change of partner. Syphilis is associated with sexual promiscuity: 'promiscuity' implies moral condemnation. But a person does not have to be promiscuous to catch syphilis. It is hard to escape the stigmatizing associations fixed to the names of some diseases. Distasteful or humiliating implications may also be strongly associated with a condition, and be the source of stigma.

People unfamiliar with Western biomedicine may try to make sense of its introduced ideas and new practices, especially when they stem from people perceived as those in authority or having power. Leprosy spread fast in the area of the West Sepik where I did field-work. With the discovery of cases, health patrols were sent. The villages do not seem to have had it before European and external contact. In the village where I lived (population 370–420) there was one person with leprosy in 1968, about 33 by 1975, and about 55 by 1985. Few diseases have such long and well-documented histories of stigma as leprosy (Brody 1974; Waxler 1981; Iliffe 1987, pp. 214–29; Lewis 1987; Gussow 1989; Jopling 1991; Vaughan 1991, pp. 77–99). In Papua New Guinea in 1968, and even in 1975, a policy of segregation for those with lepromatous leprosy was still officially in force. The segregation policy blighted a few people's lives. Leprosy did not look serious to the Gnau. People did not die from it. The signs were just marks on the skin (most of the cases among them were of tuberculoid leprosy); it was a mild (*wuyinda*) skin condition; it did not hurt or incapacitate. The people did not believe what the health education patrols said. The pictures of disfiguring leprosy they were shown did not match the signs that anyone in the village had. The skin lesions they saw looked much like other skin troubles with which they were long familiar: *gapati watelila* and *gaduet wanu'en*, which were both benign fungal infections, tinea circinata and pityriasis versicolor. Leprosy (Pigin English: *sik lepro*) did not even itch, hurt, smell or scale off as did *grile* or tinea imbricata, another quite common fungal infection. Scabies, infected sores, abscesses, and wounds could be much worse.

The swelling (disfiguring contagious) kind of leprosy (as they understood lepromatous leprosy to be) seemed different. But they expanded the category, looking for something fit for such warnings, and they applied the diagnosis *sik lepro* to other mysterious swellings that came painlessly and bloated someone's face or limbs or belly—nephrotic syndrome in one child, ascites in a man—and sometimes brought death: these seemed to justify the fuss made by the leprosy health education patrols. These must be, they thought, examples of the bad kind of *sik lepro*, the kind that could lead to people being segregated and sent for isolation. They had grasped the seriousness of the diagnosis, the idea of risk, contagion, that there were special horrors to leprosy. They were not convinced that any of that applied to the mild skin blemishes which were all that most of the affected people showed. But they extended special attitudes—fear of contagion, repugnance—to those who had signs they mistook for those of bad *sik lepro*.

The word 'leprosy' (*sik lepro*) stayed relatively fixed: what it referred to changed. The attitudes (segregation, repugnance) remained with the name. There is an echo of the Bible in this. The Hebrew word for the skin condition was translated in the Septuagint as *lepra* (deriving from the Greek *lepros*, rough and scaly, *lepis*, a scale); later its reference shifted. The skin condition described in Leviticus was not modern leprosy; Leviticus contains nothing about deformity, loss of feeling, destructive changes, or paresis. Various skin conditions might fit some of the Biblical criteria. In Leviticus, the skin condition is taken as a sign of ritual impurity, it was called unclean, the person with it was excluded from the community. It came to be associated with the idea of sin. It is hard for us now to detach the word 'leprosy' from the notion of it being a kind of disease. But it is at least arguable that the detailed Old Testament instructions to the priest on how to identify the condition of uncleanness are not about illness and medical contagion, even less about treatment in the sense of healing, but rather are concerned with religious impurity—uncleanness. We medicalize the Biblical text—we recast what may not have been considered a medical matter and discuss it as a response to perceived disease. Another curious aspect of this story is how strongly our attitudes and beliefs about leprosy are fixed. We associate leprosy with disfiguring illness and contagion. The word blinds us. Countless people have read the Biblical text without comment on the striking discrepancies between the modern disease and the 'leprosy' which is minutely described there (so the priest can identify which are the unclean lesions). That is another example of the power of social beliefs and attitudes to illness to influence perception.

CONCLUSION

AIDS in Africa: perceptions of the problems

Many of the issues discussed earlier appear sharply in the problems associated with AIDS (acquired immune deficiency syndrome): those of stigma and recognition, the classification of disease, the social origins of ideas about the illness, the social construction of attitudes to it, shifts of gaze and understanding, a broader social perspective on the causes and aetiology of the disease. I shall conclude with a review of these.

As Gallo and Montagnier (1988) remarked, 'AIDS shattered some of the confidence felt in the developed world that infectious disease was no longer really such a threat'. The swiftness of developments is frightening: HIV (human immunodeficiency virus) began to spread in the late 1970s, and it is only about 10 years ago that the first AIDS cases were recognized. The virus causing the syndrome was identified in 1983 and the outlines of the epidemic are growing clearer. The scale of the problem begins to emerge—a global problem but one in which the modes of spread are not the same all over the world. Recognition of AIDS in Africa produced a burst of questions: the prevalence was high in some central, eastern, and southern African countries; women were about as often infected as men, but in North America and Europe it had been seen at first as being a disease primarily affecting homosexual men and drug abusers. The findings from Africa provoked questions about the origins of the disease, the reasons for its prevalence, its patterns of spread, the possibility of prevention and care. It was categorized too rapidly as being a disease of the 'impure other'. It is difficult to overcome stereotypes and prejudice, for people to realize that the disease is not restricted to those with a particular sexual orientation or Africans, blacks and drug users.

For most people in the industrialized West, concepts concerning the nature of AIDS come from a mixture of sources: hearsay, news items, health education messages, television, posters. Relatively few know infected people or will have had to deal with them when they were ill. That may change. Yet we are already sharply aware of the illness and the threat. Our representations of the disease are strongly tied to its associations: sex, death, homosexuality, drugs, and blood. The emergent image is clearer as a threat than as a picture of the form that the illness might take. AIDS, as Wallman remarked (1988, p. 571), 'is an idea as much as it is a biomedical phenomenon . . . It is a powerful idea . . . associated with the perversion of both the vital fluids: blood and semen are normally the source of life; infected with an AIDS virus, they become agents of death. In addition, it is powerful because the breakdown of the immune system's capacity to inhibit infection, which AIDS causes, is a ready metaphor for the breakdown of social and sexual inhibition, which is widely believed to have caused AIDS in the first place'. This is certainly a telling instance of the social construction of illness.

By contrast, in parts of sub-Saharan Africa, infection with HIV (human immunodeficiency virus) is prevalent. Adults die young from AIDS or AIDS-related conditions, but the people in some countries were not aware of the particular nature of the disease; they had no image of AIDS to fit their experience to. They lacked cultural representations of it, despite their close experience. The epidemiologist so highly conscious of the looming dangers of AIDS in Africa may not have seen clinical AIDS first-hand or even perhaps an African village. In terms of experience, the epidemiologist is far from the events of illness, whereas a rural African may be near, indeed directly involved in them, even though he or she has no name for the illness and gives it no special recognition or distinctive explanation. The implications for finding cases and identifying the distribution of infection are obvious. The scale and distribution of AIDS in different countries is uncertain; there is widespread under-estimation and sometimes there has been deliberate neglect or suppression of information about the disease (Barnett and Blaikie 1992, pp. 1–39).

I have commented at some length on the general importance of classification and the recognition of distinctive patterns of illness. AIDS is a conspicuous illustration of this: the syndrome (and later the virus) was identified because of an unusual cluster of occurrences of some rare pneumonias and rare skin cancers; a link was made when the sufferers were noticed to be young homosexual men. The virus was identified in 1983. A new technology of blood tests quickly emerged. The forms of illness, AIDS and the AIDS-related conditions (ARC), were better identified. HIV infection could be linked to the appearance of illness. The new tests made it possible to tell whether someone was infected with the virus even if he or she felt and looked well.

The technology made possible a shift of gaze from AIDS to HIV infection. Without the tests, it was not possible to identify who was infected until he or she fell ill; and the virus is remarkable for lying latent for years before the illness develops. Over the past 10 years, knowledge of the latent period, and indeed of many other features of the disease, has changed. Breakdown into illness can come in various ways, the illness can take different shapes or disguises because opportunistic infections take advantage of the body's weakened defences. It is hard to see the connections, especially when infection is largely invisible and silent and may last for 10 years until the final stage is reached and the disease declared. There is almost nothing evident to ordinary experience to link the different conditions and features which are brought together in current medical concepts of AIDS. These concepts of AIDS are recent, they are social and medical constructions of illness and they are still changing. They heighten the contrast between the epidemiologist's perception of the problems of HIV infection and the rural African's experience of AIDS, yet lack of recognition of it.

Until recently, few rural Africans could have seen the problems in anything like the same way. They did not have the information. They had no reason to associate the illnesses and the risks or to isolate the elements which medical research now brings forward and links together. In an area of high prevalence, Rakai District, south-west Uganda, one form of illness—fever, diarrhoea, wasting to death—was sufficiently distinctive to be recognized as something new and sinister, and set apart with a name: 'slim'. Hooper (1990) has described how recognition and the name appeared in Rakai in 1985. The illness had to be sufficiently distinctive to be seen as something new. In a village, even with quite high levels of sero-prevalence (12 per cent in adults), the changes in mortality are perhaps not likely to be obvious unless people have been alerted by a question or warnings. Before the spread of AIDS, perhaps about 12 or 15 people would die in a year out of a village population of 500, and of these, 8 or 10 might come from the ranks of the very young and the old. If AIDS doubled the number of younger adults (20–45 years old) dying prematurely, that might add five or six to the annual death toll. The social effects—the drawn-out nature of the illness, the need for help with farmwork, the children left fatherless or motherless who would have to be cared for—might make the problems more apparent at the village level rather than anything distinctive or new about the kinds of illness. A survey of the village looking for clinical signs of AIDS (fever, diarrhoea, swollen glands, weight loss, persistent dry cough, itching, rashes, oral sores, genital ulcers) might reveal two or three people with oral thrush, two with herpes zoster, a few more with itching rash or swollen glands, perhaps one person seriously ill with weight loss, diarrhoea, fever, and cough, or with the serious skin lesions of Kaposi's sarcoma. The signs would not be remarkable or connected unless you knew what you were looking for. On the other hand, the number of orphans in the village grows and the problems of broken families accumulate, young widows with small children and no resources, the number of families who have lost one or more of their young adult members increases. However, perception of the situation has changed rapidly in Uganda, partly as a result of experience in the areas most seriously affected; by 1991, 20 per cent of households in some rural villages in Rakai were either directly afflicted because a family member had the disease or had died from it, or affected because death outside the family deprived them of remittances or brought responsibilities for orphans from other households (Barnett and Blaikie 1992, pp. 86–109).

It has also changed because of policy. The Ugandan government under Museveni openly declared the existence of AIDS in the country, the desperate need to control it and prevent spread. The warnings and preventive campaigns created high levels of awareness and fear, identified and gave a name and more definite shape to the increasingly bitter Ugandan experience. Other governments have been ready to hide the prevalence of AIDS, to neglect reporting it or instituting education or preventive measures. There was resentment by some African countries that they were being blamed for spread of the disease, made scapegoats, identified as its source, stigmatized as being sexually promiscuous, and morally loose. There were fears that travel between countries would be restricted, the West would use AIDS as a reason to bar them entry, blood tests would be used as controls, and people refused visas. The rapid focus of medical and epidemiological attention on African countries made some fear that their people would be used for experiments and trials from which they would derive no benefit. Indeed, there were some currents of suspicion that the 'plague' had been introduced into Africa from Europe or North America to allow experiments to be carried out for the benefit of the industrialized countries threatened with their 'gay plague'. Recognition of AIDS has changed very rapidly.

Recognition is bound up with classification and diagnosis in a search for meaning. The meaning of illness may be something that people look for in personal and moral terms, or they may interpret it through explanation of its causes. Here too the disastrous effects of AIDS in rural African communities posed special problems for explanation: young adults died in what should have been the most productive years of their lives. The destruction of normal expectations brought questions of the cause, whether it was natural or not: was witchcraft behind it? The sorts of explanation people bring forward depend on their assumptions and on the knowledge current in their society. People apply their ideas to try to make sense of the situation. In some of the worst affected areas, at first it seemed that rich traders and their wives were conspicuous among those who died. Witchcraft against them from resentment was plausible in terms of local assumptions; some of them were involved in smuggling, dealing, and getting rich at the expense of others. Health education and experience increased, and great stress was put on the dangers of sexual spread and the need for 'zero grazing' (keeping to one partner). The health education messages brought a mixture of moral and medical explanation which still left room for the element of chance or bad luck: witchcraft or sorcery could still explain why one particular person got the disease and not his neighbour who was no better or worse than he was. But the idiom of witchcraft has been replaced increasingly by ideas that take account of medical explanations (Barnett and Blaikie 1992, pp. 39–54).

AIDS is irretrievably tied to sex in almost everyone's mind: it is a sexually transmitted disease (STD). What Africa presented first was AIDS as a heterosexual, not a homosexual, illness. Medical investigations of AIDS in Africa have particularly focused on the epidemiology of heterosexual transmission of the infection. African populations have been taken as examples to picture what might happen on a far wider, perhaps a global, scale.

To characterize a disease as STD comes close to using a social behavioural characteristic to identify it. It is almost bound to have moral overtones. Epidemiological questions about the efficiency of transmission from male to female, from female to male, exact data on the population dynamics of transmission, which would depend on identifying the network of sexual partners and partner change, records of intercourse with details of sexual frequency and practice and with whom. These questions then followed by demands for blood samples, clinical, and genital examinations, obviously risk provoking extreme resentment on the part of those who are subjected to the questions. They are so narrowly focused on sex, they may seem unwarranted prying and embarrassing to answer. The investigations are also concerned with identifying risk—the times, places, occupations, and behaviours—and the identification of risk would contribute to a better understanding of cause in the broader sense. It has obvious potential relevance for planning and targeting efforts at prevention.

It is hardly surprising that research on AIDS and HIV has focused on sexual behaviour. AIDS has prompted very searching re-examination of the data recorded by anthropologists and their analyses of marriage and the values associated with sexuality and fertility (Caldwell *et al.* 1989; Larson 1990). Caldwell *et al.* (1989), in particular, have tried to set the attitudes and patterns of sexual behaviour in a coherent social framework. The distinctive African pattern (see also Goody and Tambiah 1973; Goody 1976; Tambiah 1989) sets high value on fertility, the importance of children for the continuity and strength of the descent group, and for meeting the need for labour. The argument is connected with the nature of African farming systems dependent on the hoe and enough hands to work the land rather than, as in Europe and Asia, the plough and access to land. The importance of fertility in the African pattern is linked to other variables, such as relative premarital freedom, early onset of sexual activity, polygyny, the unequal economic and domestic status of women, their roles in farm work, descent ties valued above conjugal ties, sometimes a measure of tolerance of sexual freedom for women, but clear jural rights over their children, not the same strict separation of licit from illicit sex, and trans-actions in respect of sexual services both within and outside marriage. But these studies have also provoked strong reactions and the criticism that they focus on sexual behaviour too narrowly and exclusively (Serawadda and Katongole-Mbidde 1990; Sanders and Sambo 1991). They generalize too simply in making the contrast of Africa with Europe and Asia, and fail to put it in a political context that would give a more relevant understanding. In the case of Uganda, the patchy distribution of areas of high AIDS prevalence needs to be understood in the context of political events and the history of the country. Its colonial period imposed political and landholding changes, cotton and coffee were developed as cash crops; new roads, a rail link, towns and commercial centres grew. After Independence, divisions in the country appeared, Amin seized power, Asian traders and businessmen were made to leave, and a period of mismanagement and terror followed, provoking resistance, civil unrest, and war. The disruption of normal economic life brought poverty, migration, insecurity, and encouraged smuggling, illegal trade, and traffic across borders by road and water. Poverty spurred it; migration, truck-stop bars and hotels on the trade routes, prostitution, soldiers and resistance fighters, marked a period of oppression and violence. Women depended on men for access to land, cash, and consumer goods. Women were traditionally responsible for food crops and domestic matters. During the Amin period, some men benefited from the illicit economic activity, but it increased the economic insecurity of women; some of them attempted to cope with this through various types of liaison, survival needs forced others into prostitution (Barnett and Blaikie 1992, pp. 68–85).

The problem of persuading people to take part in large-scale research studies is great. I have seen this first-hand in Masaka District in south-west Uganda, even when, as there, local people are only too bitterly aware of AIDS. The research team hold village gatherings and meetings to explain their aims, the explanatory speeches are made by respected local men and women. Village people ask why they should be made the subjects of the research. Who will benefit? Shouldn't they be paid if others benefit? Isn't the research just playing around with them? What is being done with their blood? Is it being stolen? Used to strengthen Whites? To spread the disease? Why are the researchers only interested in HIV? The villager continues with questions: Why should he give blood? What for? How much will they take? How will he or his family benefit? His mother has high blood pressure and a sore leg; what can they do for the pain in her leg? Did they bring any treatment? No? Will they be using the same needle on him as they have used for someone else? For local people, benefit from the study would first be looked for in terms of treatment. They do not want to be told nothing can be done for AIDS (if nothing, then the research is just playing around with them). They need improved health services, there are obvious crying needs for drugs and dressings, nurses and clinics. AIDS and HIV belong in a context of many other illnesses and considerable unmet basic needs for better health care.

The villagers say: 'Why are you only interested in HIV?' It is difficult to convey to them the relevance and purpose of the studies and to justify them. They look for some local benefit. In the science paradigm, data are objective; they are collected as 'facts' by observation and experiment; the geologist studies rocks and the data he obtains do not depend on the rocks being aware that they are objects of study. The scientist decides what should be studied; the objects of his investigation are passive. If social studies were able to proceed in exactly the same way as geology, the people being studied would play no part in deciding what should be investigated, how the data should be interpreted, what should be done with the results. The scientists are powerful, the objects of their inquiries weak. But in social and medical studies, the balance of power shifts. A variety of people with different interests must be recognized: the scientists with their questions, their scientific aims and career goals; the local people or the sick who have their needs and wishes, and their rights; the sponsors who fund the inquiries and have certain expectations of results; the public authorities who control access to the population, who grant research visas, foreign entry permits; the ethical committee. Barnes (1979), whose discussion of the ethics of social science research I have followed here, distinguishes them as the scientists, the citizens, the sponsors, and the gatekeepers. In the natural sciences, ethical issues arise mainly in connection with the practical application of results, but in social and medical research, they arise at the stage of data collection as well. People's interests and obligations must be taken into account. The interests of the scientists and their sponsors, and the government wanting information, must be balanced against local people's rights. The research may involve encroaching on people's privacy, the results may affect the freedom of an individual, may provide ammunition for those in power to use in justification of measures of control or restriction. The knowledge may have an effect on policy and on local people's lives; it is an element in the construction of social reality. Such knowledge is not neutral in effect.

Research on disease has often sought to keep biological matters strictly separate from social ones. But in actual illness they are not. The Gambian example with which this chapter began was about the ecology of disease and traced out some ramifying links between social causes and the occurrence of disease. The choice of relevant social facts in that example was largely decided by biomedical assumptions. But people of other cultures have different ideas. They consider other things more relevant, and their ideas are bound to influence their behaviour in health and illness. Their views about the means of knowing and the questions shift our gaze. Instead of adopting an external observer's view of the relevant facts and the connections between them, we can try to understand how they see illness. The idea of social constructions of illness contrasts with the idea of a single biological classification of disease, objective and universally valid. It suggests we should consider the cultural specificity of ideas of illness and subjective views. Behaviour and decisions about prevention and treatment follow from them. I described the Gnau and their identification of what is relevant for interpreting and explaining illness. Links between illness and social behaviour create possibilities for moral judgements to come in. Illness often acquires moral meaning. Stigma can be tied to diagnoses as the examples of leprosy and AIDS have shown. Ethical problems in research are greatly complicated by the human and moral meaning of illness. It is scarcely possible, and certainly unwise, to shut one's eyes to the reality of the mixture and confusion of biological and social facts, of values, behavious, and beliefs, in illness and in medicine.

REFERENCES

Ackerknecht, E. (1971). *Medicine and ethnology*. Hans Huber Bern, Stuttgart.
Barnes, J. (1979). *Who should know what?: Social science, privacy and ethics*. Penguin, Harmondsworth.

Barnett, T. and Blaikie, P. (1992). *AIDS in Africa*. Belhaven Press, London.

Barrell, R. A. E. and Rowland, M. G. M. (1979). Infant foods as a potential source of diarrhoeal illness in rural West Africa. *Transactions of the Royal Society of Tropical Medicine and Hygiene*, **73**, 85–90.

Beckerleg, S. (forthcoming). Food-bowl etiquette.

Brody, S. N. (1974). *The disease of the soul*. Cornell University Press, Ithaca.

Caldwell, J., Caldwell, P., and Quiggin, P. (1989). The social context of AIDS in Sub-Saharan Africa. *Population and Development Review*, **15**, 185–234.

Carney, J. (1988). Struggles over crop rights and labour within contract farming households in a Gambian irrigated rice project. *Journal of Peasant Studies*, **15**, 334–49.

Collingwood, R. G. (1945). *The idea of nature*. Oxford University Press.

Eisenberg, L. (1977). Disease and illness. *Culture, Medicine and Psychiatry*, **1**, 9–23.

Fabrega, H. (1974). *Disease and social behavior*. MIT Press, Massachusetts and London.

Fabrega, H. and Zucker, M. (1977). Comparisons of illness episodes in a pluralistic setting. *Psychosomatic Medicine*, **39**, 325–43.

Field, M. (1960). *The search for security*. Faber, London.

Frake, C. (1961). The diagnosis of disease among the Subanun of Mindanao. *American Anthropology*, **63**, 113–32.

Freidson, E. (1972). *Profession of medicine*. Dodd, Mead, New York.

Gallo, R. and Montagnier, L. (1988). AIDS in 1988. *Scientific American*, **259**, 24–39.

Good, B. J. (1977). The heart of what's the matter. *Culture, Medicine and Psychiatry*, **1**, 25–58.

Goody, J. (1976). *Production and reproduction*. Cambridge University Press.

Goody, J. and Tambiah, S. (1973). *Bridewealth and dowry*. Cambridge University Press.

Gussow, Z. (1989). *Leprosy, racism and public health*. Westview Press, Colorado.

Hooper, E. (1990). *Slim: A reporter's own story of AIDS in East Africa*. Bodley Head, London.

Iliffe, J. (1987). *The African poor*. Cambridge University Press.

Jopling, W. H. (1991). Leprosy stigma. *Leprosy Review*, **62**, 1–12.

Kennedy, J. G. (1973). Cultural psychiatry. In *Handbook of social and cultural anthropology* (ed. J. Honigmann), pp. 1119–98. Rand McNally, Chicago.

Kleinman, A. (1980). *Patients and healers in the context of culture*. University of California Press, Berkeley and Los Angeles.

Kleinman, A. and Good, B. (ed.) (1985). *Culture and depression*. University of California Press, Berkeley and Los Angeles.

Kunitz, S. (1987). Explanations and ideology of mortality patterns. *Population and Development Review*, **13**, 379–408.

Lamb, W. H., Foord, F., Lamb, C., and Whitehead, R. (1984). Changes in maternal and child mortality rates in three isolated Gambian villages over ten years. *Lancet*, **ii**, 912–14.

Larson, A. (1990). The social epidemiology of Africa's AIDS epidemic. *African Affairs*, **89**, 5–25.

Lawrence, F., Lamb, W., Lamb, C., and Lawrence, M. (1985). A quantification of childcare and infant care-giver interaction in a West African village. *Early Human Development*, **12**, 71–80.

Lewis, G. (1975). *Knowledge of illness in a Sepik society*. Athlone Press, London.

Lewis, G. (1980). *Day of shining red*. Cambridge University Press.

Lewis, G. (1987). A lesson from Leviticus. *Man*, **22**, 593–612.

Littlewood, R. and Lipsedge, M. (1982). *Aliens and alienists*. Penguin, Harmondsworth.

Lock, M. and Gordon, D. (ed.) (1988). *Biomedicine examined*. Kluwer, Dordrecht.

McGregor, I. A. (1990). The origins of Keneba MRC laboratories in the Gambia. *The Biologist*, **37**, 43–6.

McGregor, I. A., Billiewicz, W. Z., and Thomson, A. M. (1961). Growth and mortality in children in an African village. *British Medical Journal*, **2**, 1661–6.

McKeown, T. (1976). *The modern rise of population*. Academic Press, New York.

McNeill, W. (1979). *Plagues and peoples*. Penguin, Harmondsworth.

Mechanic, D. (1962). The concept of illness behavior. *Journal of Chronic Diseases*, **15**, 189–94.

Mechanic, D. (1968). *Medical sociology*. Free Press, New York.

Nichter, M. (1989). *Anthropology and international health*. Kluwer, Dordrecht.

Obeyesekere, G. (1976). The impact of Ayurvvedic ideas on the culture and the individual in Sri Lanka. In *Asian medical systems* (ed. C. Leslie), pp. 201–26. University of California Press, Berkeley and Los Angeles.

Parsons, T. (1951). *The social system*. Free Press, New York.

Prentice, A. and Prentice, A. (1988). Reproduction against the odds. *New Scientist*, **1608**, 42–6.

Reiser, S. J. (1978). *Medicine and the reign of technology*. Cambridge University Press.

Roberts, S., Paul, A., Cole, T., and Whitehead, R. (1982). Seasonal changes in activity birthweight and lactational performance in rural Gambian women. *Transactions of the Royal Society of Tropical Medicine and Hygiene*, **76**, 668–78.

Rowland, M. G. M., *et al.* (1981). Seasonality and the growth of infants in a Gambian village. In *Seasonal dimensions of rural poverty* (ed. R. Chambers, R. Longhurst, and A. Pacey), pp. 164–92. F. Pinter, London.

Sanders, D. and Sambo, A. (1991). AIDS in Africa. *Health Policy and Planning*, **6**, 157–65.

Serawadda, X. and Katongole-Mbidde (1990). Viewpoint: AIDS in Africa, problems for research and researchers. *Lancet*, **335**, 842–3.

Simons, R. and Hughes, C. (ed.) (1985). *Culture-bound syndromes*. Reidel, Dordrecht.

Sydenham, T. (1742). *The entire works of Dr Thomas Sydenham* (ed. J. Swan). Edward Cave, London.

Szreter, S. (1988). The impact of social intervention in Britain's mortality decline. *Social History of Medicine*, **1**, 1–38.

Tambiah, S. (1989). Bridewealth and dowry revisited. *Current Anthropology*, **30**, 413–32.

Topley, M. (1970). Chinese traditional ideas and the treatment of disease. *Man*, 5, 421–37.

Vaughan, M. (1991). *Curing their ills*. Polity Press, Cambridge.

Wallman, S. (1988). Sex and death: the AIDS crisis in social and cultural context. *Journal of Acquired Immune Deficiency Syndrome*, 1, 571–8.

Waxler, N. (1981). Learning to be a leper. In *Social contexts of health* (ed. E. Mischler *et al.*), pp. 169–94. Cambridge University Press.

Weaver, L. and Campbell, F. (1989). *Nutrition in a developing world: the work of the Dunn Nutrition Unit in the Gambia 1988–1989*. Medical Research Council, Cambridge.

Whitehead, R. (1985). *Dunn Nutrition Unit: Cambridge, UK and Keneba, The Gambia 1982–85*. Medical Research Council, Cambridge.

Whitehead, R. (1988). *Keneba update 1986–1987*. Medical Research Council, Cambridge.

Wilcocks, C. (1962). *Aspects of medical investigations in Africa*. Oxford University Press.

Williams, R. (1976). *Keywords*. Fontana, London.

Wrigley, E. and Schofield, R. (1981). *The population history of England 1531–1871*. Edward Arnold, London.

Young, A. (1976). Internalizing and externalizing medical belief systems. *Social Science and Medicine*, 10, 147–56.

Young, A. (1982). The anthropologies of illness and sickness. *Annual Reviews of Anthropology*, 11, 257–83.

Zborowski, M. (1958). Cultural components in responses to pain. In *Patients, physicians and illness* (ed. E. G. Jaco), pp. 256–68. Free Press, New York.

3

HISTORICAL AND CONTEMPORARY MORTALITY PATTERNS IN POLYNESIA

Stephen J. Kunitz

INTRODUCTION

In this chapter I shall consider the impact that different patterns of European contact have had on the evolution of mortality and epidemiological regimes in Polynesia. In particular, I shall use historical and contemporary demographic and epidemiological data to address several issues having to do with the various ways in which people have attempted to understand social and economic change and their consequences for the well-being of populations.

Polynesia is as close to an ideal place as the observational scientist can find to examine the consequences of different patterns of European contact because the biological and cultural similarities among people scattered across thousands of miles of ocean and different island environments have been obvious to observers since the first European contact.[1] Perhaps unique in the world, Polynesia thus lends itself to the method of controlled comparison, which is as close as the observational scientist can come to an experimental situation (Eggan 1954; Oliver 1981).

The first issue I shall deal with has to do with the impact of European contact on the trajectory of indigenous populations. For more than a century there has been a conviction that the peoples of the Pacific experienced major losses at least since the time of first European contact (e.g. Rivers 1922; Hamlin 1932; Lambert 1934). Some have contended that the decline began even prior to contact (Roberts 1927). Explanations have varied. Epidemics introduced into 'virgin soil' populations by European explorers and colonists are common to virtually all of them. Declining fertility as a result of declining *joie de vivre* (Rivers 1922), or as a result of the disruption of traditional social organization and culture (Pitt-Rivers 1927), or as a result of venereal diseases, has also been implicated.

More recent writers have criticized these older views. Vern Carroll (1975) has called this received history a 'myth' which may have been true of some atoll populations but certainly not of all. Norma McArthur (1968) and Ian Pool (1977) have criticized both the inadequate population data used by previous writers as well as the explanations of the causes of population decline they offered, particularly those that were based upon various versions of psychological or psychoanalytical theory. Further, K. R. Howe (1984, pp. 347–52) has argued that the impact of European contact was not as fatal as many have said, and that the notion that Pacific peoples were dying out was largely self-serving and used to justify colonial domination: as the natives were dying anyway, the best that could be done was to smooth the death-bed pillow. Indeed, this was one of the arguments used to justify annexation of Hawaii by the United States (Heffernan 1988, p. 223). The 'new' Pacific history, Howe writes, views the indigenous populations as far more resilient, resistant, and adaptive than previous observers had thought. For example, he has claimed that the Maori population in 1769 was essentially the same size as in 1840. This assertion illustrates the well-known difficulty encountered by all who are interested in population change: estimating the number of people around the time of first contact from archaeological data and impressions of early explorers. Clearly, the size of the population at contact influences whether we assess the impact of the contact experience as fatal or benign. This, in turn, has important political implications, for the greater the collapse the more pressing are the moral claims for reparations to contemporary indigenous people (e.g. Stannard 1989, p. 143).

The second issue I shall address has to do with the problems that arise when we attempt to explain contemporary disease patterns by the degree to which people are said to be 'traditional' or 'modern'. As illustrated in the so-called Mead/Freeman debate about Samoa, the very nature of traditional society is often far from clear and a source of bitter disagreement (Mead 1928, 1930; Freeman 1983; Holmes 1987). In part, the lack of clarity may result from differences among investigators, and the periods and places they have studied; all of which have been offered as partial explanations of the different accounts of Samoan society. In part, lack of clarity may have to do with assumptions about the course and definition of 'modernization', and with how we understand the persistence of what appear to be traditional forms of social organization and cultural values and their relevance to health and well-being.

The third issue has to do with how similar social structural arrangements may lead to different endpoints. I shall consider the differences between Maoris and Native Hawaiians in contemporary life expectancies to make the point that even within the so-called Fourth World of indigenous peoples overwhelmed by European contact there continue to be important distinctions which influence rates of mortality. The fourth issue has to do with the different ways in which very different social and economic arrangements may lead to the same epidemiological endpoint.

In each instance I shall use comparisons among and within societies to make the point that the demographic and epidemiological patterns that have been reported are explainable in terms of social organization, patterns of European contact, and social and economic change. I shall deal only with some of the large islands because small atoll populations are extremely vulnerable to a wide variety of random (i.e. unexplained) influences as well as to environmental influences, such as hurricanes and drought, to which the large islands are more resistant (Vayda 1959; McArthur *et al.* 1976).

HISTORICAL PATTERNS OF POPULATION CHANGE

Figure 3.1 shows estimates of Polynesian populations from five large island groups: Hawaii, New Zealand, Tonga, Western and American Samoa combined, and Tahiti and the Marquesas in French Polynesia.[2] Figure 3.2 plots the same data on a semi-log scale in order to compare rates of change. Although recognizing the very imperfect nature of the data, several points suggest themselves.

First, sustained population decline was not a universal phenomenon. It is widely agreed that there were major losses of the Hawaiian population beginning in the 1780s (Schmitt 1968; Nordyke 1989), but the actual numbers are a matter of considerable debate. Estimates of the population at contact range from 250 000 to almost a million. I have used the presently more widely accepted lower figure. Warfare, epidemics, and sub-fecundity due to the spread of venereal diseases have all been blamed. Certainly, the rapid growth of Honolulu from a small settlement in the early nineteenth century to 6000 by the 1830s would have resulted in the efficient dissemination of a variety of diseases throughout the population (Daws 1967; Ralston 1977, pp. 101–3).

Likewise, the New Zealand Maori population declined from the time of first contact, but recent estimates by Pool (personal communication) suggest that the major losses occurred after the 1840s when contact with Europeans intensified. His revised estimate of the population at contact in 1769 is now 110 000 rather than 125 000–175 000.

The population of Tahiti also declined, but the range of estimates of the contact population (16 000–50 000) is very large. I have arbitrarily used the middle of the range in my plots, but even accepting the lowest estimate would reveal a very substantial loss. By the 1820s, however, population

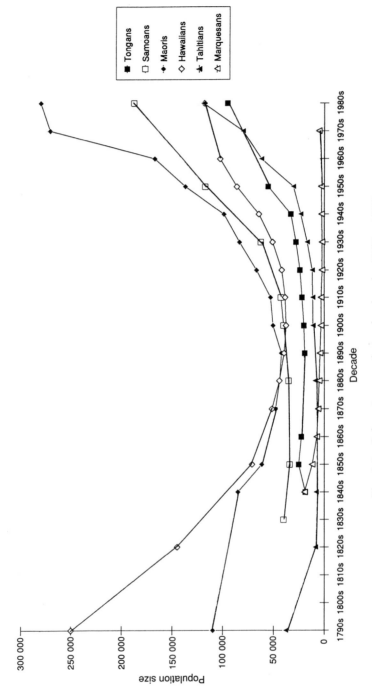

Fig. 3.1 Polynesian population by decade, 1790s–1980s.

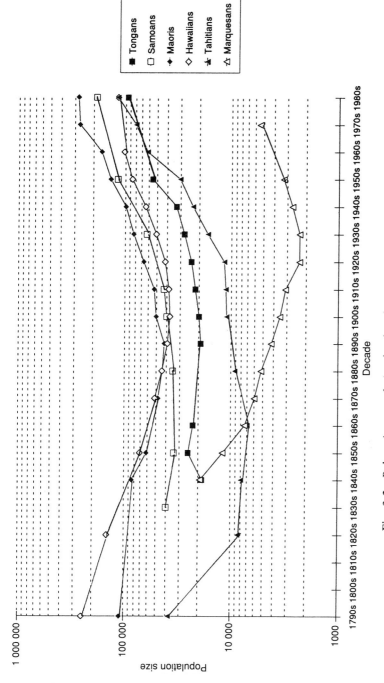

Fig. 3.2 Polynesian population by decade, 1790s–1980s, semi-log scale.

appears to have stabilized. Similary, the Marquesas lost a large part of its population, but unlike Tahiti decline continued right into this century.

On the other hand, Samoa and Tonga seem to have experienced little or no decline from the first estimates in the 1830s and 1840s. It is of course possible that they lost population earlier, but at least for Samoa there seems to be no evidence for this (C. Macpherson and L. Macpherson 1990, pp. 54–8). Indeed, their population trajectories in the nineteenth century are very much like Tahiti's after 1820, whereas New Zealand's and Hawaii's native populations declined into the 1890s. No matter what their nineteenth-century histories were, however, all these populations but the Marquesans began to increase dramatically around the turn of the present century.

How are we to explain these patterns? It is useful to distinguish the early period of contact in the late eighteenth and early nineteenth centuries from later periods: the second two-thirds of the nineteenth century, say, and the present century. In the earliest period massive population loss seems to have occurred in Hawaii and Tahiti, and probably in the Marquesas (Rallu 1990, pp. 48–9). New Zealand suffered a somewhat lower rate of loss; and Samoa and Tonga seem to have lost little or no population.

What evidence there is suggests that the intensity of contact with Europeans largely accounts for these differences. The quest for firearms provoked conflict (e.g. Newbury 1980, pp. 10–11), and their acquisition made conflict more lethal; but this was a widespread phenomenon and occurred in Samoa as well as Hawaii. The frequency with which vessels called seems to have varied considerably, however, and early in the nineteenth century was greatest in Tahiti, Hawaii, and New Zealand. On the other hand, the very large size and low population density (see Table 3.1) of New Zealand would appear to have protected much of the population from contact until settlement intensified several decades later.

The more readily answerable as well as to me more interesting question has to do with the differences in population patterns from about the 1830s onwards. Hawaiian, Maori, and Marquesan population declined; Samoan Tongan, and Tahitian population stagnated but did not decline substantially, if at all. Setting aside consideration of the Marquesas for the moment, the most striking difference between Hawaii and New Zealand on the one hand and the Samoas, Tonga, and Tahiti on the other is that during the nineteenth century the former were settled by Europeans and Americans who dispossessed and demographically overwhelmed the indigenous peoples, whereas the latter became colonial outposts with a small European population attempting to extract resources from the numerically superior indigenous peoples. That is to say, the latter were much more nearly colonies in the usual way the term is understood. The former are examples of what Donald Denoon (1983) has called 'settler capitalism', a term he has applied to South Africa, Argentina, Uruguay, Chile, and Australia as well as New Zealand.

As he describes it, 'settler capitalism' has several characteristics which distinguish it from other versions of colonialism (Denoon 1983, pp. 221–4). (1) Each began as a garrison outpost of one or another European empire. (2) 'There was no dependable production, because there was no exploitable indigenous community strong enough to sustain a stratum of conquering settlers.' (3) Pastoralism usually dominated production at first; landowners consolidated control over land and labour 'while a benign administration registered their titles and protected their property.' (4) During the nineteenth century 'these settler societies took full advantage of new production and transport and market opportunities, to achieve a level of prosperity and a demographic and territorial expansion which none had imagined in 1814.' The settlers' 'independence was not complete, but it was much more substantial than that which prevailed in most of the tropical world, where conquest and colonial administration prevailed.' (5) Yet diversification of production did not occur. Britain influenced the quantity and quality of production through market conditions rather than direct imperial control. (6) Such 'unforced dependence' was the result of powerful internal forces (social classes) whose interests were compatible with British imperial interests. (7) As settler societies expanded they came in contact with agricultural populations which, under certain circumstances, entered into cash crop production and became peasantries. This was a transient phase, however, and was quickly transformed. 'In general, the effect of settlers upon other rural people was to drive them swiftly towards fully capitalist relations of production, passing briskly through peasantization and plantation production merely as transitional social formations.'

Although not included in Denoon's analysis, much of Hawaii's history may be described in very similar terms. Most significant for present purposes is the final point, that the agricultural peoples encountered by the settlers were within a short period turned into wage workers, either rural or urban. This was as true of Hawaiians as of Maoris. In both cases, settler control of land led to expropriation and the consolidation of large tracts for ranching or (in Hawaii) plantations which employed either the indigenous people or, if they could not be coaxed or coerced or were not numerous enough, by imported contract labour. Similar attempts were made in what became Western Samoa and Tahiti, but for reasons which I shall return to below such attempts were not as successful as in New Zealand and Hawaii.

The economic transformations of these settler societies attracted highly mobile and landless Europeans who were adaptable to the conditions of these new (to them) lands (Denoon 1983, p. 217). This European (and, in Hawaii, Asian) demographic wave which numerically and economically overwhelmed Maoris and Native Hawaiians was partly responsible for the catastrophic declines each experienced. It is usually argued that exotic diseases were the cause of the decline experienced by these and other virgin soil populations. That such diseases as influenza, measles, and tuberculosis were of enormous significance cannot be doubted. On the other hand, such diseases were also introduced into the Samoas, Tonga, and Tahiti with serious but not cataclysmic consequences.

Others have observed that European contact in Western Polynesia (Samoa, Tonga) has had less impact on population than in Eastern Polynesia, although these islands were not free of exotic epidemics (C. Macpherson and L. Macpherson 1990, pp. 54–8). Pirie (1972, pp. 198–9) has suggested several possible reasons. (1) Western Polynesia may have been less isolated biologically than Eastern, so at contact the population was less vulnerable to the Europeans' diseases. (2) Contact with Europeans may have been regulated more formally by chiefly authority in Western Polynesia than elsewhere. (3) A lower prevalence of yaws in Eastern Polynesia may have produced less cross-immunity to syphilis than may have been the case in Western Polynesia.

Without disputing the possible significance of any of these factors, it seems most likely to me that social disruption caused by the kind of settler capitalism I have already described is likely to prove a more significant part of the explanation. I am not invoking a psychosomatic explanation in which longing for the lost home results in depression, suppression of the immune response, and increased vulnerability to disease, although all of that is possible. Rather, it seems likely that the impoverishment resulting from the destruction of subsistence agriculture would have made people more susceptible to respiratory diseases and gastroenteritis, which flourish under conditions of poverty, crowding, and malnutrition. Moreover, the expropriation of land resulted in removal and very likely in the disruption of social networks which provided both instrumental and emotional support in times of need. Observations of epidemics in virgin soil populations suggest that social disruption is at least as significant in causing high mortality as is the virulence of the infectious agent itself, and many contemporary studies suggest an important role for social support in reducing the risk of death from a wide variety of causes.

Moreover, the greater number of Europeans in New Zealand and Hawaii than in Tahiti, Samoa, and Tonga would have increased the risk of sexual contact and of the spread of sexually transmitted diseases with a probable decrease in fecundity as a result.

On the other hand, in the more tropical islands the transformation of the indigenous peoples into wage workers did not occur as swiftly or as thoroughly. The traditional economy was 'peasantized', to use Denoon's term. That is to say, the indigenous peoples engaged in both subsistence activities and in cash cropping of bananas, copra, and taro. Although attempts were made to create plantations on both Samoa and Tahiti, this never accounted for most of the suitable land or workforce as it did in Hawaii and New Zealand. It is useful to ask: Why?

It is unlikely that patterns of chiefly centralization and decentralization are entirely responsible. Hawaii was highly centralized under Kamehameha I at the beginning of the nineteenth century, but so was Tahiti under Pomare, and Tonga under Taufa'ahau. New Zealand was highly decentralized, as were the Marquesas and Samoa. It is often argued that land alienation was not permitted by the Tongan king, but it also true that Tonga was not sufficiently attractive to Europeans to make any foreign power wish to expend the effort to expropriate the land (Scarr 1990, p. 219).

Indeed, it is the amount of arable land that seems to best account for differences in settlement between New Zealand and Hawaii on the one hand, and Samoa, Tonga, and French Polynesia on the other. Table 3.1 displays the area of each group of islands. Because all except Tonga are mountainous, the amount of land suitable for agriculture is, of course, much less than the total area. Unfortunately, data on the amount of suitable agricultural land are not readily available. Crocombe (1987, p. 20) has commented concerning land policy 'The goals of centralized [colonial] administration were to obtain effective control, reduce dispute, increase production, and *in the larger territories*, to make way for colonists' (emphasis added). Missionaries and colonial administrators often aspired to make producers of the natives—either on their lands or on plantations—and to keep out large numbers of settlers (Scarr 1990, p. 264). This seems to have been true on Tonga and Samoa (e.g. Meleisea 1987, p. 79) and in French Polynesia (Newbury 1980, p. 112). They were more likely to succeed on islands which were not attractive to large numbers of settlers and thus where they could retain some control and where massive land alienation did not occur. Moreover, in the few cases where plantations were

TABLE 3.1. Areas and populations of selected Polynesian islands

Island	Area (km^2)	Population/km^2 at first estimate*
Tonga	699	26.4 (1840s)
Western Samoa	2934	
American Samoa	197	
Total	3131	12.7 (1830s)
French Polynesia	3265	
Society Islands	1626	9.8–30.7 (1790s)
Hawaii	15 862	15.7
New Zealand	269 063	0.4

* See Appendix.

established on Samoa and Tahiti, it was very difficult to get the indigenous people to work on them because they preferred to work their own land as peasant producers.

Every bit as striking as the different patterns of decline and stability during the nineteenth century is the similarity in patterns of growth in the twentieth century. Beginning in the second and third decades of this century there was a significant increase in all populations, again excepting the Marquesans. There is no evidence of radically improved and widespread economic well-being through the 1940s when these changes were getting under way. Nor is there yet evidence of some sort of genetic selection for disease resistance among these populations. There is, however, evidence of attempts to improve public health on the part of the indigenous people themselves, colonial and indigenous governments, and non-governmental organizations, particularly the Rockefeller Foundation (Lambert 1941). It is not possible to assess the impact of any of these programmes: one can simply point to the temporal association between such attempts and the increase in population.

First, in respect of self-help, the 'Maori renaissance' of the first years of this century was a movement in which a few Maori leaders began to engage in a variety of preventive and health educational activities. Ian Pool (1977, pp. 110, 144) has suggested that this movement may have been responsible for, or at least contributed to, the 'turnaround' in Maori mortality at this time, although firm data are lacking. Another example is the development of women's committees in Samoan villages in the 1920s. These were stimulated originally by a New Zealand physician and involved influential women in each village who exercised surveillance of 'child care and village health' (Keesing 1945, pp. 221–2). These committees continue to operate in many villages, although again with effects that do not seem to have been measured (Schoeffel 1984).

In respect of government action, quarantine was of course an important function. Just how important is illustrated by the fact that the influenza epidemic of 1918 was prevented in American Samoa, which enforced quarantine, but killed perhaps 20 per cent of the population in Western Samoa where the New Zealand administration failed to enforce it (Meleisea 1987, p. 121). The distribution of free food for infants was a function of the government of Tonga (Keesing 1945, p. 219). Sanitary regulations were passed by most governments, although they seem to have been honoured more in the breach than in the observance in many places (Keesing 1945, p. 221). The protection of water supplies was also attempted in a variety of regions.

In respect of the Rockefeller Foundation's activities, this was originally part of the world-wide hookworm eradication campaign, but also involved treatment and prevention of yaws. Because hookworm is spread by human faecal contamination of the ground, the Foundation's work on waste disposal may have reduced the spread of enteric diseases as well. As in the preceding examples, there is no way of assessing the magnitude of the impact these measures had on the mortality of the population. It seems likely, however, that the similar trajectories of population growth after the 1910s is indeed the result primarily of public health measures which were widespread in these years.

The point I have made in this section is two-fold. First, the picture of population collapse painted by early writers was partly true, but not universally. It was not true of most island populations where American and European settlement was not extensive. It was catastrophic in those populations where settlement and dispossession occurred. This is important because it suggests that the contact situation mediated between the new microorganisms to which people were exposed and the mortality that resulted. The experience was not everywhere the same, but the differences are largely understandable.

Even this statement needs to be qualified, however, by the experience of the population of the Marquesas, which declined throughout the nineteenth century and most of this century as well. Until the 1920s, the Marquesan death-rate was substantially greater than the birth-rate, and life expectancy at birth was only 21.5 in 1886–1910, and 17.4 in 1911–25; increasing to 34.5 in 1925–45 (Rallu 1990, p. 176). The same factors invoked to explain the early declines in population have been offered for the Marquesan situation: venereal disease causing reduced fertility; extreme maternal mortality; and high death-rates from introduced diseases, particularly tuberculosis and other respiratory conditions, and diseases of infancy (Rallu 1990, pp. 154–5). What is not clear is why this situation persisted so long among the Marquesans but not elsewhere because they were not overwhelmed and dispossessed as were the Maoris and Hawaiians. Many observers have noted, however, that violence and conflict were a continuing feature of Marquesan life long after they had ceased elsewhere in Polynesia (e.g. Newbury 1980, pp. 142, 197), and presumably this social disruption—thought by many to be a function of Marquesan culture itself—contributed substantially to the high mortality rates.

Life expectancy at birth and age 10 improved more slowly in the south-eastern than in the north-western islands of the Marquesas group, from the 1880s to the 1940s (Rallu 1990, p. 163). In 1936–45 life expectancy at birth was 49.2 years in the north-western islands population and 37.6 in the south-eastern islands population. By way of comparison, in 1944–46 Maori life expectancy was 48.4, and in 1940 Native Hawaiian life expectancy was about 52 (see Table 3.4 below). These data suggest that in the south-eastern islands living conditions were significantly worse than elsewhere and that endemic diseases affected both children and adults long after they had improved elsewhere. Unfortunately, the reasons remain obscure.

My second point has been that widespread population growth was recorded from the 1910s onward, again excepting the Marquesas where improvement began two or three decades later. I have suggested tentatively that the most likely explanation is the fairly broad array of public health interventions that were introduced at this time.

The distinction between settler capitalist (or 'settler colonial') and colonial society largely explains the different population trajectories of the nineteenth century. It does not explain the similarity of their trajectories in this century. Nonetheless, the distinction continues to be relevant in the present to the degree that the socio-economic, cultural, and political circumstances of Maoris and Native Hawaiians are similar to one another and different from the circumstances of the other populations. To explore some of these contemporary differences and similarities, I have displayed in Table 3.2 life expectancies at birth around 1980 for females and males in each of the populations under consideration.

In the following sections I shall use the contrast between Western Samoa and American Samoa to explore what we mean by the distinction between 'traditional' and 'modern' as they influence health. I shall compare New Zealand and Hawaii to explore how similar social structural situations may lead to different life expectancies. And I shall compare American Samoa and Hawaii to explore how different social policies may lead to similar life expectancies.

TABLE 3.2. Life expectancy at birth of several Polynesian populations, c.1980

Population	Males	Females
Western Samoa (1)	60.6	66.1
American Samoa (1)	67.8	75.5
Tonga (1)	60.8	65.2
French Polynesia (1)	60.1	64.2
New Zealand Maoris (2)	63.8	68.5
Native Hawaiians (3)	70.8	76.0

(1) Taylor *et al.* (1989). (2) Pomare and de Boer (1988). (3) Gardner (1984).

HEALTH CONSEQUENCES OF TRADITIONAL AND MODERN LIFE

It is in the work of John Cassel that ideas about the health consequences of socio-cultural change first most explicitly entered the field of epidemiology (Cassel et al. 1960). Drawing heavily on Robert Redfield's (1955) notion of the folk–urban continuum, he characterized traditional folk societies as well integrated, with a coherent moral order on which all members agreed. Social relationships were face-to-face and stable; and people were known in many different roles, not simply in their roles as occupational specialists or as family members and friends.

Cassel described modern urban society as individualistic. People know one another only in special contexts. There is pluralism of religious, ethical, and political beliefs. Values and social relationships change rapidly. Once adjusted to, he said, urban life is not necessarily stressful and damaging to health. But the transition from traditional to modern often is, because people are not equipped socially or psychologically to cope with the new situations with which they are faced.

These ideas have informed much subsequent research, including an important and very valuable study of the health of Samoans to which I shall refer frequently in the following discussion. The point I wish to make is this. Notions of folk and urban, or traditional and modern, which are used to explain the higher prevalence of certain non-infectious conditions in modern societies are generally studied synchronically but are assumed to represent temporal change. The problem is that the communities that are said to be traditional do in fact depend heavily on subsistence activities similar to those practised in past times, but the whole context in which these activities occur is so changed that to describe the villages as traditional is potentially misleading. The studies of the health of Samoans are a useful illustration of this problem.

I have grouped Samoa with those colonies whose history contrasts with the settler capitalist societies which developed in New Zealand and Hawaii. Of course, after 1899 there were two Samoas, American Samoa and Western Samoa, which came under the control of different colonial powers with very different goals. The American government assumed control over the island of Tutuila and Manu'a Islands. The German government assumed control of the islands of Upolu and Savai'i.

The Americans wanted the deepwater harbour at Pago Pago on Tutuila for a mid-Pacific coaling station. The island is mountainous and not suitable for extensive agriculture, and there was no intention to turn it into a colony for the extraction of primary products. The Manu'a group 60 miles to the east was even less suitable. The Navy administered the islands with what appears to have been a relatively light hand until World War II, when large numbers of servicemen were stationed there and when large numbers of Samoan men got jobs around the naval base. This was the time when wage work and cash entered the local economy on a very large scale. After the war, administration was taken over by the Department of the Interior, to which the local government is responsible (Sunia 1983).

The Germans had very different goals for the two islands they acquired, which had already become the centre of Germany's expanding Pacific empire. The firm of Godeffroy und Sohn, which had been in Samoa since the mid-nineteenth century, and for which the German consul worked, acquired large tracts for plantations. In the late nineteenth century these assets were acquired by Deutsche Handel und Plantagen Gesellschaft, which also worked closely with the German administrator, Wilhelm Solf. Solf resisted attempts of settlers to acquire property, believing that, 'German racial and economic interests as well as Samoan interests, were best served by a company-operated large-scale plantation economy' (Meleisea 1987, p. 79). New Zealand assumed control over Upolu and Savai'i during World War I. The islands gained independence as the nation of Western Samoa in 1962. These very different histories have had profound consequences for the people of the two Samoas which may be summarized briefly.

In a study of seven villages on Tutuila in American Samoa and one on Savai'i in Western Samoa in 1986, Fitzgerald and Howard (1990) found that educational, income, and employment levels were all lower in Western Samoa, and mean household size was larger. Individuals in Western and American Samoa were equally likely to get cash from family members, but the amount received was greater in American Samoa, and money was given to a wider network of kin. Although equal proportions of individuals in each place reported serving a *matai* (the chiefly head of a family), 'In Western Samoa 91.7% of matai served were within the respondent's household, while in American Samoa only 23.9% were in the household' (p. 45). This seems to reflect differences in household size but as Fitzgerald and Howard indicate, raises questions about the role of chieftainship in each place as well. Moreover, in Western Samoa a higher proportion than in American Samoa report giving daily service to a *matai* (See Table 3.3).

American Samoa also has a much higher proportion of its population employed by the government, and in 1990 had about as many physicians (34 providing patient care) for an estimated 46 000 people as Western Samoa did for an estimated 160 000 people (McCuddin 1989; Department of Health 1990, pp. 60–1). Physicians were paid a great deal more in American Samoa as well. Starting salary in 1991 was about US$42 000; not a great deal by mainland American standards but considerably more than the range paid in Western Samoa: WS$10 785–18 080 (approximately

TABLE 3.3. Socio-economic measures in American and Western Samoa

	American Samoa	Western Samoa
Education (yrs) (1)		
Males	12.4	8.9
Females	13.0	9.8
Employed (%) (1)		
Males	61.5	34.8
Females	60.9	32.0
Mean annual income (US$) (1)		
Males	3256	642
Females	3969	315
Household size (1)	7.7–7.9	11.5–13.1
Residence of *matai* within household (%) (1)	23.9	91.7
Daily service to *matai* (%) (1)	18.4	89.6
Total population	29 301 (2)	157 408 (3)
Proportion of pop. > 15 yrs employed by government	24.7 (2)	4.2 (3)
Total population, per capita income (US$)	3144 (2)	712 (4)

(1) Fitzgerald and Howard (1990): survey data are from seven villages in American Samoa, one in Western Samoa.
(2) Bureau of the Census (1984): per capita income is from 1979.
(3) Department of Statistics (1988): data are for 1986.
(4) O'Meara (1990, p. 189): data are from the early 1980s.

US$5000–9000). The result has been the migration of eight Western Samoan physicians to American Samoa and the vacancy of a number of positions in the Western Samoan health service. The same sort of labour migration occurs at all occupational levels, of course. It is said that 80 per cent of the cannery workers in American Samoa are from Western Samoa.

At the level of territory or nation, then, the story seems rather straightforward: higher income and greater availability of health services in American Samoa have resulted in substantially greater life expectancy there than in Western Samoa. If, however, one considers not the mortality rates and life expectancy but the prevalence of certain non-infectious diseases, the picture looks somewhat different; because as Cassel's theory leads one to expect, their prevalence and incidence are higher in urban than rural areas.

The conditions that have been found to increase as the infectious diseases wane are non-infectious and man-made in origin (Omran 1971; Baker and Crews 1986; Crews 1988a, b, 1989; Crews and Pearson 1988). They are attributed to changes associated with 'modernization', particularly changes in consumption and activity patterns and increasing psychosocial stress. I shall deal with each briefly, using non-insulin-dependent diabetes to exemplify the first and hypertension the second.

A large number of studies of non-insulin-dependent diabetes in Pacific Island populations have been done over the past two decades. They all indicate that prevalence is higher in urban than in rural populations. Indeed, Polynesians and Micronesians, like Australian Aborigines and North American Indians, tend to have very high rates when compared to many other peoples, and it is widely thought that an as yet unknown genetic predisposition must be an important part of the explanation, along with changes in diet and exercise.

Obesity, for instance, is a well known risk factor for non-insulin-dependent diabetes. A number of studies among Samoans show that urban residents (in the Samoas as well as in the United States and New Zealand) often become massively obese, whereas rural residents, especially men, generally do not. Part of the explanation has to do with the pattern of feasting on Sundays among both urban and rural people, unaccompanied by strenuous exercise among the urban residents to utilize the extra calories (Hanna et al. 1986, p. 296). Among rural men doing agricultural work, the problem does not arise. Among women, who tend to be sedentary whether urban or rural, obesity is a common condition. Obesity alone does not explain the high prevalence of diabetes, and it has been proposed that some of the unexplained variance in the prevalence of diabetes might be the result of 'stress' (Zimmet et al. 1981). I shall return to this issue below.

A roughly comparable but somewhat more complex picture emerges from studies of conditions that are generally thought to be stress-related, particularly elevated blood pressure. Studies of blood pressure are frequent both because it is widely thought to be responsive to psychosocial conditions and because it is easy and inexpensive to measure. Usually, differences in pressure are analysed and treated as an interval variable rather than normotension or hypertension being considered a dichotomous variable. The point is important because statistically significant differences in blood pressure using tests of interval data may or may not reflect differences that are of substantive (i.e. clinical) significance. Hypertension is an important risk factor for premature death, in Samoans as well as other people (Crews 1988a), whereas an elevation of 4–5 mmHg may not be.

John Cassel (1974) wrote that in traditional societies hypertension is infrequent and does not increase with age, whereas in modern societies it is more frequent and does increase with age. Hypertension (defined as a systolic blood pressure >160 mmHg or a diastolic pressure >95 mmHg) increases with age in rural and urban Samoan populations. Moreover, although the numbers are small in several of the samples, there is no consistent difference in the prevalence of hypertension across communities at ages 55 and above (McGarvey and Schendel 1986, p. 374). At ages below 55, however, there is a strong tendency for populations in rural areas to have lower prevalence rates of hypertension then populations in urban areas. Because these are cross-sectional studies, it is impossible to know how much of the observed patterns are accounted for by migration.

When level of blood pressure is considered rather than a diagnosis of hypertension, the data suggest that people experiencing some sort of stress have significantly higher levels. In the study under discussion: 'Stress is viewed as the result of a disjunction between demands of the social environment and an individual's coping resources or ability to solve problems' (Howard 1986, p. 395).

Assessing the contribution of psychosocial processes to increases in blood pressure is difficult because the most important determinants of elevated blood pressure are increased weight, age, and salt intake, and because the definition and assessment of processes, such as stress, are far from straightforward, particularly cross-culturally. There is, however, some suggestive evidence that occupational demands on men educationally ill-equipped to meet them are associated with elevated blood pressure. Men in Pago Pago, American Samoa, with less than seven years of education and occupying managerial positions had higher levels than people whose positions were consistent with their educational attainments (McGarvey and Schendel 1986, p. 380). The same result is not observed among women. It has also been observed that men with low educational attainments who do not express emotional complaints (on the Cornell Medical Index) have higher blood pressures than those who do, suggesting—but not demonstrating—that denial of emotional difficulties may be associated with increases in blood pressure.

In a study of Samoans in California, Janes (1990, pp. 118–23) also explored the relationship between social inconsistency and blood pressure. He observed that two dimensions were especially important within the Samoan community: economic status and leadership in the church or family. Outside the Samoan community he considered the same variables as had been used in the study in Pago Pago: occupational status and education (or military service). Inconsistency was said to be present when leadership roles and economic status did not match (i.e when status was high and income low, or vice versa), and when occupation and education did not match. In both cases systolic blood pressures of men were significantly higher among those whose positions were inconsistent than among those whose positions were consistent. Diastolic pressures did not differ; nor were there differences among women. On the other hand, for women problems related to family life were associated with increases in both systolic and diastolic pressures (Janes 1990, pp. 124–5). Janes (p. 123) wrote of his results:

The very characteristics that make Samoan leadership status a goal that men pursue avidly are those that render its attainment a powerful stressor. Seen in this way, social inconsistency is probably as much a feature of traditional Samoan society as it is of the stateside community. However, where wage labor and status in the wage economy become added prerequisites to the attainment of leadership status, the potential for inconsistency likely increases.

This is an important observation. No studies have investigated similar relationships in rural Samoan communities where lower average blood pressures and lower rates of obesity prevail. It does suggest, as Howard (1986) has argued, that the conditions of urban life are more diverse than those of rural life, and that individuals' repertoires of skills and resources must also be more diverse in urban communities as well.

The question is whether in the Samoan context it is helpful to explain these conditions as resulting from differences in traditional and modern society. A reasonable case can be made for doing so: rural villagers engage in activities and adhere to forms of social organization more nearly like those of their ancestors than do urban dwellers. The Samoan health study defined modernization as follows (Hecht et al. 1986, p. 41):

Several processes of culture change are associated: increasing universality of education, commercialization and monetization, development of communication and transportation links, increasing concentrations of population, the general availability of formal health services, and the expansion of alternatives to the extended family and kin networks. In the Samoan project studies, 'modernization' is used as a rubric for these co-occurring processes without implications of causality.

Although it is clear that these characteristics are indeed often found together, it may be misleading to claim that if a community or population does not have these attributes it is traditional. It may just be poor. This is an important point, for traditional implies that such communities and populations are not part of the modern world, indeed that like the mythical village of Brigadoon they are largely untouched by it. The evidence in respect of Samoan villages suggests something quite different: that agriculturists have sought actively to participate in the cash economy and to bring improvements to their villages when they could afford them. Schoeffel (1984, pp. 211–12) has observed that: 'One of the ironies of modern Samoan history is that although the Samoans themselves look back upon the colonial period as one in which Samoans tenaciously opposed European cultural influences and European defined notions of economic change, the rapid pace of modernization in Western Samoa since independence suggests that colonial paternalism rather than Samoan traditionalism was the real conservative force in the past.'

O'Meara (1990) in his fine-grained analysis of the economic and social life of an agricultural village in Western Samoa supports this observation.

He has made the important point that the villagers responded in the 1950s and 1960s to increasing banana and cocoa prices by increasing production, selling their produce, and using the cash for such co-operative self-help projects as electrification of the village. When commodity prices collapsed, the booms ended. 'Today, lacking a viable internal source of income, they rely on external gifts. In many ways their village is more backward now [in the 1980s] that it was a generation ago' (p. 67).

Indeed, he discovered that among the 56 village families (virtually the entire population) the single largest source of income was gifts from overseas. Of a total cash income of WS$120 000 during one year in the early 1980s, almost WS$31 000, or just over 25 per cent, came from this source. The sale of coconuts—mainly in the form of copra—amounted to almost as much (WS$27 800). Earned income totalled WS$19 800; small family-run businesses brought in WS$14 000; the sale of taro WS$13 100; and gifts from people in other villages WS$11 600. He estimated that the monetary value of the agricultural produce grown in the village for domestic consumption just about equalled the income received from the sources listed above: WS$120 000 (O'Meara 1990, pp. 184–9).

There has been continuing pressure from development specialists to get farmers to do more cash cropping and to change the land tenure system from family to individually owned plots. O'Meara argues that the land tenure system has undergone dramatic change over the past 70 years: 'Yet the agricultural revolution has not followed. From this we see that traditional Samoan social institutions are not blocking development.' 'The major obstacle', he continues, 'lies elsewhere—in the economics of village agriculture' (p. 162).

In general, village planters have lower incomes than the Western Samoan nation-wide average. Within the village, farm families earn far less than families who receive cash from abroad, from wage work, or from small businesses. 'Even among the . . . farm households, sales of agricultural products account for less than one-third of their monetary income' (p. 189). Income per day of labour is lower in agricultural than in non-agricultural pursuits, and the marginal return to agricultural labour is lower than in wage work as well. The economic incentives for cash cropping have been 'woefully inadequate', he concludes, so villagers have turned elsewhere in their search for money (O'Meara 1990, pp. 189–92).

In the light of these data, I should like to suggest that the continued presence of subsistence agriculture and extended networks of kin are not simply evidence of the persistence of a traditional society and culture, but an effective adaptation to the modern world in which the collapse of commodity prices has reduced the ability of farmers to participate profitably in the world market. This does not mean that the results of numerous studies showing differences in the health of rural and urban people are wrong. It means, rather, that the attribution of the causes of the differences to modernity and traditionalism may be misleading because it assumes unlinear social evolution, whereas what we may be seeing instead is two very different patterns of adaptation to two very different forms of colonialism and social and economic change.

The point is that there are real changes in health associated with the move from rural to urban life. Not only do the infectious diseases decline, but some non-infectious conditions increase. Most of the increase has to do with changes in food consumption and activity; some of it has to do with changes in social organization and with the diversification of the roles that people have to play. Because subsistence agriculture continues to be a viable adaptive strategy in the contemporary economic environment of rural villages, forms of behaviour and social organization that have been part of that adaptation also continue to be viable, although there is also evidence of profound changes (e.g. C. Macpherson and L. Macpherson 1987). This is not as true in cities, where the demands on individuals have become increasingly diverse, and where their repertoires of psychosocial and cognitive skills must also diversify. The problem that has concerned me here has been the use of the labels 'traditional' and 'modern' to explain these differences, for they imply not simply evolutionary development from one to the other, but an ahistorical conception of the conditions under which 'traditional' behaviours remain viable in the modern world. It is for that reason that I have suggested that tradition and poverty have been conflated.

DISTINCTIONS WITHIN THE FOURTH WORLD

In his 1939 monograph, *Some modern Hawaiians*, the New Zealand anthropologist Ernest Beaglehole drew a number of contrasts between Maoris and Hawaiians. Hawaiians had much less experience as farmers than Maoris, he wrote. Of 107 Hawaiians applying for homesteads on Molokai, for instance, 12 were farmers. The rest were 'stevedores, carpenters, engineers, mechanics, clerks, firemen, mail carriers, with further scattering members of professional or semi-professional occupations' (Beaglehole 1939, p. 44). Although recognizing the potential bias in the sample, he argued that the Maoris' situation was very different (p. 44):

The Maori has always been a rural people, working for himself or for the white man on farms and ranches. Once rehabilitation [the consolidation of fragmented land holdings] was initiated, the Maori, as an experienced farmer, rurally minded, well-trained in the techniques of farm life, took easily to a way of life that was merely the continuation under different conditions of an occupation with which he was familiar and at which he was well trained. By virtue of tribal discipline, enlightened native leadership and a feeling for occupational continuity, the Maori has been relatively successful where his northern Polynesian cousin, the Hawaiian, has required, and still requires, years of paternalistic support, generous financial help, and an intensive economic education before a successful citizen-farmer movement can be considered accomplished.

The fact that the Hawaiian homesteaders were given inadequate agricultural land, the best having been taken by the sugar plantations (Howard 1974, p. 4), was not mentioned by Beaglehole as an important part of the problem. Nonetheless, he was pointing to a real difference in the experience of the two peoples. With respect to contemporary leadership among Hawaiians and Maoris, he made the following observation (Beaglehole 1939, pp. 100–1):

My last [Hawaiian] informant made a pertinent comparison between the lack of Hawaiian leadership and the success attained by Maori leaders in New Zealand, of which he had heard from visitors. The comparison emphasizes a valid distinction between the history and cultural attitudes of the two Polynesian peoples. The Maori tribe, the effective unit of social and political organization, never disintegrated under white pressure in any manner comparable to the breakdown of Hawaiian political organization. Tribal loyalty is still intense and consequently tribal leadership is still respected and followed. Where leadership in modern economic schemes has passed to those whose prestige is not hereditary in the tribe, this leadership receives full support from hereditary chiefs . . .
The disappearance of the chiefly class in Hawaii and the lack of a traditionally validated dependence of the individual upon the integrated blood group symbolized by its chief have left the modern Hawaiian leaderless at the very time when native leadership might well have formulated and pursued a vigorous policy of cultural conservation.

He went on to say that: 'Dominant and inspired leadership on the one hand, a fiercely flaming feeling for tribal and cultural integrity on the other hand—the union of these two produced a Maori renaissance. He would be a bold observer indeed, who saw either of these two characteristics in modern Hawaii, or would prophesy their birth in the immediate or distant future.'[3]
The difference in political leadership resulted from the decentralized nature of Maori social organization, which contrasted strongly with the highly centralized form that Hawaiian organization took when Kamehameha gained power after the coming of the Europeans. The intermarriage of the Hawaiian royal family with the newcomers, and the loss of the land base in which chiefly status was rooted, worked to create an egalitarianism which permeates contemporary Hawaiian culture (Howard 1974, pp. 27–8).
The Hawaiian and Maori experiences have also differed, however, as a result of differences in the economic development of New Zealand and Hawaii. As a result of the growth of the plantation system in Hawaii in the

second half of the nineteenth century and the first several decades of this century, labourers from several different nationalities were recruited sequentially: Chinese, Japanese, Portuguese, and Filipinos most numerous among them. Native Hawaiians intermarried with all these groups to a considerable degree. Moreover, as Hawaiians were succeeded as plantation workers by other nationalities, and as the best agricultural lands were taken by the plantations, an increasing number of Native Hawaiians moved to Honolulu (and later Hilo) in search of work (Lind 1938, p. 325). For example, the proportion of Hawaiians and part-Hawaiians who were urban residents increased from 30.4 to 45.8 per cent from 1900 to 1920 (Bureau of the Census 1922).[4] The process was accelerated as job losses occurred on plantations in the 1930s. The populations of the counties of Maui, Kauai, and Hawaii actually declined while the population of the county of Honolulu continued to increase rapidly (Nordyke 1989, pp. 102–5).[5]

The residential mixing of races, although it had of course occurred in the nineteenth century as well, was greatly accelerated by urbanization in the 1920s and 1930s (Lind 1938, pp. 308–13). On plantations racial groups had lived in segregated settlements. In Honolulu ethnic neighbourhoods were established but mixing became increasingly common. 'This has been most noticeably true in the case of the Hawaiians and part-Hawaiians', wrote Andrew Lind (1938, p. 311), 'whose diffusion throughout the city is now unquestionably the greatest of all the ethnic groups in the city. There remains only one small section of the city where the Hawaiians and part-Hawaiians constitute the dominant group, while in 1900 one-third of all the city tracts were so characterized'.

The decline of ethnic or racial neighbourhoods—what Lind (1930, 1931) regarded as the transition from ghetto to slum—led, he claimed, to the mixing of races, the breakdown of institutions of social control, 'deculturization', and an increase in deviant behaviour. He wrote: 'For the city as a whole we find a rough inverse correlation between social disorganization, measured in terms of juvenile delinquency and dependency, and the degree of segregation and concentration of the immigrant colony' (Lind 1931, p. 210). By the 1930s, then, contemporary observers were arguing that Native Hawaiians had become increasingly urban, increasingly scattered among other ethnic groups, and increasingly distant from their own culture. The process continued after World War II, and by 1980 almost 80 per cent of Native Hawaiians lived in 'urban' places (population more than 2500), the majority in Honolulu.

In contrast, Maoris during the early decades of the century were still overwhelmingly rural, as Beaglehole's (1939) observations suggested. In 1936 about 90 per cent of the Maori population lived in primarily rural counties (McCreary 1968, p. 202), and it was the belief of many Maoris as well as non-Maoris that they were better off there. (Hawaiians believed the same thing about their situation, Howard 1974, pp. 3–4.) Indeed, land settlement schemes were meant to make it possible for them to remain an agricultural population (Ngata 1940). It was clear to some observers even then, however, that this would not be possible because population had been increasing rapidly for two generations by that time and soon there would be neither land nor employment to support them in rural areas (Belshaw 1940). Urban migration was inevitable.

Indeed, in the post-war decades Maori urban migration increased substantially. From about 9 per cent in urban areas in 1936, the proportion increased to almost 15 per cent in 1945, and about 29 per cent in 1961 (McCreary 1968, p. 200). By 1986, the proportion of Maoris in urban areas (small towns and main centres) was about 80 per cent (Pomare and de Boer 1988, pp. 28–9; Pearson 1990, p. 111). Thus, although at present approximately equal proportions of Hawaiians and Maoris live in urban areas, the process of urbanization began about 40 years earlier among the former than the latter.

These differences in rates of urbanization are related to differences in the economic histories of Hawaii and New Zealand. I have already indicated that plantations ceased to be the single most dominant force in Hawaii's economic life in the 1930s. Since World War II the tourist industry has emerged as a major employer, with Hawaiians generally working at the lowest levels of the occupational hierarchy (Kent 1983). Over the same period, New Zealand's economy has also shifted from producing agricultural goods for export to the production by protected industries of manufactured for the domestic market (Pearson 1990, p. 112). As in the case of Native Hawaiians, so too in the case of Maoris there is over-representation at the lower ends of occupational and economic hierarchy.

In general, the American (and Hawaiian) economy has been more ebullient than New Zealand's, although New Zealand has historically devoted a larger proportion of its resources to health and welfare benefits for its citizens. According to the World Bank, 1986 GNP per capita in the United States was US$17 480, whereas in New Zealand it was US$7460. In Hawaii in 1983 the comparable figure was US$12 100 (Garwood 1986); presumably it was slightly higher in 1986. I have estimated Maori per capita income in 1986 to have been about US$3600 (derived from Pearson 1990, p. 135). Per capita income of Native Hawaiians in 1979 was US$5328 (Bureau of the Census 1988). These data, and particularly those for the Maoris, are not precise, nor do give an adequate picture of the inequalities of income distribution within each group. They do, however, give a broad picture of the magnitude of the income differences between them.

These historical and contemporary differences in urbanization and income are associated with historical and contemporary mortality patterns. In Table 3.4 is shown data on life expectancy at birth for most of this

TABLE 3.4. Life expectancy at birth, Maoris and Hawaiians, twentieth century

Date	Maoris (1)		Hawaiians (2)	
	Male	Female	Male	Female
1910	32.5*		30.2	30.4
1920	na	na	35.9	34.2
1925–27	49.4–46.6	49.6–44.7	na	na
1930	na	na	42.2	43.7
1935–37	46.3–48.8	46.0–48.0	na	na
1940	na	na	51.0	53.8
1944–46	48.8	48.0	na	na
1950	54.1	55.9	61.3	64.0
1960	59.1	61.4	63.0	67.0
1970	61.0	65.0	65.0	69.9
1980	63.8	68.5	70.9	76.0

na: not available.   *Combined life expectancy.

(1) *Sources:* Pool (1985): p. 232 for 1910; Pool (1977): p. 154 for the 1920s to 1960s; Pomare and de Boer (1988): p. 31 for the 1970s to 1980s; the figures for 1950, 1960, 1970, and 1980 for the Maoris are in reality 1950–52, 1960–62, 1970–72, and 1980–82.

(2) *Sources:* Gardner (1980): p. 222 for the 1910s to 1970s; Gardner (1984): p. 13 for 1980. Gardner writes (1980, p. 222) that the 1910 figures 'should be viewed skeptically'.

century for Maoris and Hawaiians. Table 3.5 shows the average annual changes in life expectancy at birth from one decade to the next.[6]

It is clear that in the 1920s and early 1930s Maori life expectancy was greater than Hawaiian life expectancy at birth. It is also clear, however, that Maori life expectancy was essentially stagnant from the mid-1920s to the mid-1940s, whereas Hawaiian life expectancy improved substantially over the same period. Life expectancy improved very dramatically in the 1940s for Hawaiians and in the 1940s and 1950s for Maoris. Improvement has continued right up to the early 1980s, the last years for which data are available. Despite the dramatic improvement of Maori life expectancy since World War II, however, their life expectancy in 1980–2 was still substantially less than that of Hawaiians. As Table 3.6 indicates, the differences are not simply due to infant and child mortality but are evident at every age. For example, Hawaiians have higher expectation of life at age 65 than Maoris do at age 60.

There are several issues to be dealt with, therefore. Why did Maoris have higher life expectancy than Hawaiians in the 1920s? Why did Hawaiian life expectancy accelerate in the 1920s to 1940s when life expectancy was stagnant among Maoris, improving only in the late 1940s? What accounts for the patterns of change in the 1950s to 1970s? And why is life expectancy

TABLE 3.5. Average annual changes in life expectancy at birth, Maoris and Hawaiians, twentieth century

Date	Maoris	Hawaiians
1920–30	na	0.79
1925–27 to 1935–37	0.0	na
1930–40	na	0.94
1935–37 to 1944–46	0.12	na
1940–50	na	1.03
1944–46 to 1950–52	1.1	na
1950–60	0.53	0.23
1960–70	0.28	0.25
1970–80	0.32	0.6

na: not available.

Source: Derived from Table 3.4; figures for females and males are combined.

TABLE 3.6. Life expectancy at various ages: Maoris and Native Hawaiians, *c.*1980

Age (yrs)	Maoris (1980–82) (1)		Hawaiians and part-Hawaiians (1980) (2)	
	Males	Females	Males	Females
0	63.8	68.5	70.9	76.0
5	na	na	67.4	72.2
15	na	na	57.6	62.4
20	46.2	50.7	na	na
25	na	na	48.6	52.8
35	na	na	39.7	43.3
40	28.1	31.9	na	na
45	na	na	31.0	34.0
55	na	na	23.1	25.4
60	13.4	16.4	na	na
65	na	na	16.2	17.5
75	na	na	10.2	10.5

na: not available.

(1) Pomare and de Boer (1988, p. 31).
(2) Gardner (1984, p. 13).

at every age now greater among Hawaiians than Maoris? The answers can only be suggestive, not definitive.

With respect to the first question, Pool (1985, p. 232) has stated that from the late nineteenth century to the 1910s and 1920s Maori life expectancy increased dramatically at the time when, 'In what would today

be termed a "primary health care" programme, the Maori medical prac-
titioners who joined the fledgling Department of Public Health attacked
more immediate risk factors such as nutrition and basic hygiene, as well as
some of the underlying aspects of Maori social and economic deprivation'
(Pool 1985, p. 234). This so-called 'Maori renaissance' may well have had
a profound impact on living conditions and health in the first several
decades of this century. Nothing comparable occurred among Native
Hawaiians at that time, as Beaglehole's (1939) observations suggest. On the
other hand, Maori life expectancy began to stagnate in the late-1920s while
Hawaiian life expectancy continued to increase rapidly. Why?

It seems likely that the explanation has to do with the impact of the
Depression on the primarily rural Maoris, which must have weakened many
public health programmes. Hawaiians, however, were moving in large
numbers to Honolulu in these same years. This migration, while un-
doubtedly creating social disruption of the sort described by Lind (1938)
and others in the 1920s and 1930s, occurred at a time when urban mortality
had for the first time become lower then rural mortality for people throughout Western Europe and North America. That is, by the late nineteenth
century public health measures were being introduced in cities and were
having a profound effect on the spread of communicable diseases, particularly those that were food- and water-borne. Immigrants to American cities
from poverty-stricken parts of Eastern, Central, and Southern Europe had
lower infant mortality rates than people who remained behind (Kunitz
1984).[7] There is no reason to think the experience of urban migration
would have been different for Native Hawaiians.

Supporting evidence comes from a comparison of changes in life
expectancy at birth of the entire Native Hawaiian population with those of
Native Hawaiians on the large island of Hawaii, which is primarily rural
except for the urbanizing area of Hilo. (See Table 3.7.) Clearly the rate of
improvement for people on the big island was substantial but much less
than it was for the total Native Hawaiian population.

The very dramatic acceleration of Maori life expectancy after World War
II cannot be adequately explained by the same kind of urbanization which
influenced Hawaiian life expectancy during the inter-war years because
Maoris remained predominantly rural until well into the 1960s. It seems
instead to have been the result of major changes in public policy. Ian Pool
(1985, p. 234) has summarized them as follows:

Decreases [in mortality] were produced by a combination of factors: (a) the
introduction in 1938 and extension over the next few years of a free health care
scheme giving equal access to hospital and other services (before 1938 health care
was on a fee-for-service basis, or by charity); (b) the slotting of this into a
comprehensive social welfare system also introduced in 1938; (c) the introduction
of new medical technology, notably antibiotics and chemotherapeutic drugs.

TABLE 3.7. Life expectancy at birth, total Native Hawaiian population, and Native Hawaiian population of the island of Hawaii, 1910–50

Date	Total Native Hawaiian population (1)	Island of Hawaii (2)
1910	30.2	29.5
1920	35.9	31.3
1930	42.2	35.5
1940	51.0	39.2
1950	61.3	45.9

(1) *Source:* Table 3.4. (2) Fleischman (1982, p. 7): figures are for 5 years centred on the census year, e.g. 1910 is 1908–12, etc.

The gains in Native Hawaiian life expectancy, which were almost as great as were the gains for the Maoris in the 1940s, seem to me almost certainly due to the availability of antibiotics as well as to the rapid post-war expansion of the economy. There were no special programmes devoted to the improvement of the health of Hawaiians of anything like the magnitude of those for the Maoris in those years. On the other hand, increasing employment in wage work was accompanied by increasing health insurance coverage which would have facilitated access to medical care. For instance, in a study in 1967 (admittedly some years after the period of most dramatic improvement), 86 per cent of a sample of Hawaiians living in a community about an hour's drive from Honolulu has some form of health insurance coverage, compared with 70 per cent of the national population at that time (Heighton 1968, p. 121).

For Native Hawaiians and Maoris the rate of improvement diminished after the 1950s as a result of both the rapid decrease in the importance of infectious diseases as a cause of death and an increase in the incidence of non-infectious conditions such as cardiovascular diseases, cancer, and diabetes. Among Maoris, data from the mid-1960s suggested that obesity, hypertension, diabetes, and cardiovascular diseases had become very prevalent. For example, the death-rate from ischaemic heart disease among Maoris aged 65 and above was 140/10 000 in 1956–60 (Prior 1968, p. 283); and 183/10 000 in 1980–84 (Pomare and de Boer 1988, p. 76). Because of the way the published data are presented, it is more difficult to detect the pattern of change for those aged 45 to 64 over this same period, but from 1975 to 1980–84, there was a decline in deaths from ischaemic heart disease. Pomare (1980, p. 23) noted there had been little change from 1968 to 1975. The age-specific death-rate from hypertensive heart disease among Maoris 65 and above was essentially unchanged from 1954–58 to 1975, and then declined from 24/10 000 to 13.2/10 000 in 1980–84. The pattern seems to have been roughly the same among people aged 45 to 64 years old (Prior 1968, pp. 283–4; Pomare 1980, pp. 21–4; Pomare and de Boer 1988, pp. 72–4).

Although the data are made somewhat murky by changes in the accuracy of diagnosis, coding convention, and diagnostic rubrics, there is some reason to believe that among middle-aged people, deaths from hypertensive and coronary heart disease increased through the 1960s and then began to decline in the 1970s. Indeed, very likely the decline in death rates from ischaemic heart disease contributed substantially to the slight increase in the years added to life expectancy from 1970 to 1980.

A similar patter has been observed among Native Hawaiians. Death-rates due to hypertensive heart disease peaked in 1950 and have declined continuously since then. Death-rates from arteriosclerotic heart disease and from diabetes peaked in the 1960s and have declined subsequently (Look 1982). The increase in the number of years added to life expectancy from 1970 to 1980 is probably due largely to these recent improvements, which parallel improvements in other populations in Western Europe, North America, New Zealand, and Australia. Thus, although there have been parallel changes in the cause-of-death structures of both populations, they seem to have been the result of somewhat different constellations of forces. Urbanization and economic improvement seem to have begun earlier for Hawaiians and to have been of relatively greater importance than specific health and welfare policies in improving life expectancy. The reverse seems to have been true for Maoris.

There is some suggestive evidence that these differences may also help explain the contemporary differences in life expectancy displayed in Table 3.6. I have already shown that Hawaiian per capita income is higher than Maori income. There is also evidence that the educational attainments of Hawaiians are greater than those of Maoris, perhaps because of their longer history of urban life. In 1981, the proportion of Maoris aged 15 and above who had finished 7th form (high school) was 18.6 per cent (Department of Statistics 1982, p. 89). In 1980 the proportion of Native Hawaiians age 15 and above who had graduated from high school was about 62 per cent (Bureau of Census 1988, p. 841). Leaving aside questions of educational quality and equivalence, the important point is that a far higher proportion of Hawaiians than Maoris had satisfied criteria that would allow them either to go on to further education or equip them with a minimum credential necessary for employment.

Higher income and educational attainment usually translate into improved health in a variety of ways: through the ability to purchase better housing and safer automobiles; through differences in exposure to occupational hazards; and through differences in behaviour. I have not found good comparative data on all these domains, but some information is available.

For example, among Maori men, mean body mass index varies from 27.9 to 35.2 depending upon age; among women it varies between 25.8 and 30.5. Among Hawaiian males, average body bass index is 26.6; among females 25.4 (Pomare and de Boer 1988, p. 150; Le Marchand and Kolonel 1989, p. 142). Maoris thus tend to be somewhat more obese than Hawaiians. Among Hawaiians, 43 per cent of males and 38 per cent of females are smokers, whereas among Maoris the comparable figures are 53.5 per cent and 58.5 per cent (Pomare and de Boer 1988, p. 153; Le Marchand and Kolonel 1989, p. 140). It is difficult to compare consumption of carbohydrates, fats, and protein because of the way the data are presented. It appears, however, that average daily intakes of fats, proteins, and cholesterol are lower among Hawaiians than among Maoris (Pomare and de Boer 1988, pp. 151–2; Le Marchand and Kolonel 1989, p. 142), and that Hawaiians' intake of saturated fats is lower and of unsaturated fats higher than Maoris'. With respect to prevalence rates of hypertension, again comparisons are very difficult due to problems of age adjustment and definitional criteria. Taking that into account, the reported rates do not appear substantially different (Pomare 1980, p. 32; Wegner 1989, p. 119). In each case, rates (including borderline cases) seem to be in the vicinity of 500/10 000 (age adjusted to the non-indigenous populations in each instance). Thus, many—but by no means all—of the measures which are considered risk factors for various non-infectious diseases seem to be elevated more among Maoris than Hawaiians.

Causes of death do not differ very dramatically, however. For example, Table 3.8 displays the data for all forms of heart disease. Data availability do not permit the same sort of comparison of death-rates from various cancers. Thus, it is not possible using published data to determine what causes of death are responsible for lower life expectancy among Maoris than Hawaiians. One can say, however, that the differences in life

TABLE 3.8. Age-specific death-rates per 10 000 population, due to all forms of heart disease, Native Hawaiians and Maoris

Population	Males (age yrs)		Females (age yrs)	
	45–64	>65	45–64	>65
Native Hawaiians (1980–86) (1)	59.3	268.8	33.3	217.9
Maoris (1980–84) (2)	67.9	285.9	34.2	204.8

(1) Wegner (1989, p. 125); (2) Pomare and de Boer (1988, pp. 74–6).

expectancy are compatible with the differences in income, education, and risk factors reported for each population.[8]

These differences between Maoris and Native Hawaiians suggest the following. The similarity in their histories of contact has resulted in very similar consequences in respect of their incorporation into the lowest economic and occupational ranks of the settler societies that engulfed them. This has meant that in each society their health status and life expectancies are less favourable than those of other groups. Understandably, it is these inequalities and the legacy of injustices of which they are a part that are the concern of the people themselves.

On the other hand, the two societies have had very different social policies. The result has been that Native Hawaiians are among the poorest of an affluent society whereas Maoris are among the poorest of a middle income society. The social policies pursued by New Zealand have ameliorated these conditions, particularly in the 1940s and 1950s when infectious diseases were predominant. In the current epidemiological regime dominated by non-infectious diseases such interventions are unlikely to have as profound an effect, even if the political commitment were as vigorous as it was four decades ago. Thus, the differences between these two populations support the conventional wisdom that it is better to be rich than poor; as well as the more contentious notion that generally it is better to be poor in a rich country than in a poor one. Finally, they lead us to ask, under what economic and epidemiological conditions can social welfare policies (including but not limited to health care) overcome low income to increase a population's health to a level comparable with that of a higher income population?

SETTLER COLONIALISM AND WELFARE STATE COLONIALISM

The question with which the previous section ended may be addressed by considering how peoples with histories as different as those of Hawaiians and American Samoans have attained such similar life expectancies. The material presented so far may be used to draw a number of contrasts between them. Clearly their nineteenth-century population histories have been profoundly different: Hawaiian numbers declined dramatically; Samoan numbers remained essentially constant. Hawaiians were deprived of land; Samoans were not. Hawaiians had sexual relations and bore children with members of other races; Samoans have not done so to nearly the same extent. Native Hawaiians and part-Hawaiians have tended to remain in Hawaii; a very high proportion of Samoans has emigrated. Per capita income in 1979 was US$5328 for Hawaiians and US$3144 for

Samoans in American Samoa. The proportion of people 15 years of age and older who were high school graduates in 1980 was 62 per cent among Hawaiians and 40 per cent among American Samoans. The proportion of people 15 years of age and above in the labour force was 63.4 per cent among Hawaiians and 45.5 per cent among American Samoans; and of these, 24 per cent and 50 per cent, respectively, were employed by a governmental agency (Bureau of the Census 1984, 1988). Thus, unemployment is higher in American Samoa than among Hawaiians in Hawaii, and a higher proportion of the employment that does exist is in government.

American Samoa is among the most extreme examples of what have been called MIRAB states. This is an acronym which has been applied to Pacific island micro-states characterized by high levels of migration, remittances from abroad, foreign aid, and bureaucracy (Bertram and Watters 1985, 1986; Hayes 1991). Aid is provided by the former colonial powers either for strategic or ethical reasons. The term 'welfare state colonialism' has been applied to this relationship, for the island micro-states are clearly dependent upon the former colonial powers, but just as clearly the situation is different from simple colonialism. In this context aid may be regarded as a form of rental income. Because it is not a loan, it does not have to be repaid; and therefore it does not lead to indebtedness. Such funds, although often given for economic development, in fact largely finance local government. Thus, government jobs are very attractive because they are stable and the major local source of cash income in many of these island economies. As my discussion of Western Samoa illustrated, the other major source of cash is remittances from relatives working abroad.

These patterns also characterize Indian reservations in the United States. Rather than strategic concerns, aid is provided to meet treaty obligations. Indeed, American Samoa is very similar to an Indian reservation. It is administered by the same federal agency (the Department of the Interior); it has the same service-based employment and economic structure (Kunitz 1983); and it has a very similar health care system, with services generally being accessible and provided at little or no cost to the patient.

If American Samoa is like an Indian reservation, however, the situation of Native Hawaiians is much like that of American Indians east of the Mississippi. These tribes were conquered during the colonial period, and therefore when treaties were made, it was with the governments of the original colonies, which subsequently became states, not with the federal government. State governments have not honoured their treaty obligations as has the federal government because they are under much more undiluted pressure from mining, ranching, forestry, and (in Hawaii) plantation and tourist industry interests to allow the natives' land to be used for the profit of others. Countervailing pressures from reform-minded urban constituencies are more likely to be able to influence the federal government than state governments. Thus, like Indians in the eastern United States, Native Hawaiians are much more integrated into the non-indigenous population than are Indians on federal reservations or than American Samoans. Also like Indians in the eastern United States, Native Hawaiians are attempting to gain recognition by the federal government of special status which would entitle them to many of the same benefits enjoyed by American Samoans and by Indians on federal reservations: the return of land, access to special health services, special schools, and education programmes, and so on (Native Hawaiians Study Commission 1983). This may be seen as an attempt to convert the dispossession characteristic of settler colonialism into the entitlements characteristic of welfare state colonialism.

These very different patterns of colonial contact have, however, resulted in rather similar life expectancies at present. The fact that per capita income in American Samoa is about 60 per cent of the per capita income of Native Hawaiians and roughly the same as Maori income suggests that it is their stable if low level of employment, the safety valve of emigration, and the high level of services (including not simply personal care but also public health programmes) that is largely responsible for the similarities. Comparison of survival curves (see Fig. 3.3) and infant mortality rates for 1980 suggest that the slightly lower life expectancy at birth of American Samoan than Hawaiian men was the result of somewhat higher death-rates at all ages, particularly in adulthood. Life expectancy of women in each population was essentially the same (Crews 1985, pp. 82–3; Gardner 1984).

The contrast between Maoris and Native Hawaiians suggested that a reasonably buoyant economy such as Hawaii's could lead to increases in life expectancy even if the absence of special entitlement programmes; and that a depressed economy such as New Zealand's in which such programmes were eroding would result in comparatively less favourable life expectancies. The contrast between Hawaiians and American Samoans suggests that generous welfare entitlements can in fact lead to high life expectancy even in the absence of a buoyant economy. Such a result may seem obvious. It is not. If it were, there would be far less disagreement about the health consequences of such programmes, although disagreement would certainly persist over their economic viability, especially if they are universalized.

CONCLUSIONS

Several points emerge from the preceding discussion. They have to do with biological, social, and economic determinism. By biological determinism I mean the argument put forth by several historians, perhaps most notably by William McNeil (1976) and Alfred Crosby (1986), and in more popular form by Alan Moorehead, that the impact of European contact on the

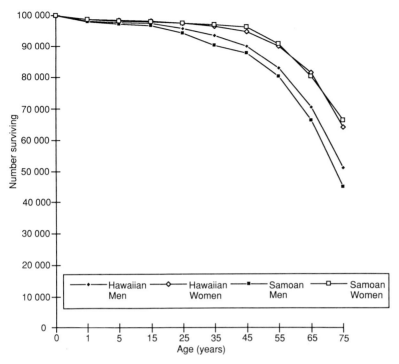

Fig. 3.3 Survival rates of Native Hawaiian and American Samoan males and females at various ages, 1980.

peoples of the Americas and Oceania was uniformly cataclysmic in respect of population size. To deny that contact led to widespread population decline is as unreasonable as denying that the Holocaust took place. Nonetheless, it is now clear that there were differences in the responses of virgin soil populations to newly introduced diseases. They did not all succumb in large numbers. Elsewhere I have suggested that some of the difference had to do with the social organization of indigenous peoples themselves: hunter–gatherers seem to have been more vulnerable than agriculturalists (Kunitz 1990). Here, I suggested that the kind of colonial contact which occurred was also of enormous importance. The significance of this observation is that it means that social forces mediated between newly introduced infectious organisms and their human hosts.

   Lest we be tempted to conclude, however, that social forces are all-important and that the science-based control of infectious diseases is largely irrelevant—what I have termed social determinism—the history of simultaneous population increase across societies with very different histories of colonial contact should prove a useful antidote. To be sure, these public health interventions were themselves social products. Nonetheless, they were broadly effective in several different contexts.

Finally, the recent history of these various societies suggests that economic development is not invariably associated with high life chances and non-development with low life chances. The relatively high life expectancies of the populations of some MIRAB states and Indian reservations suggests that heavily subsidized non-development may be an equally effective way to improve the health and well-being of certain peoples: those whose lands are too remote, too small, and too poor in natural resources to ever be truly 'developed'. This, of course, means a transfer of wealth from rich to poor, and it is thus only in special cases—e.g. for strategic considerations, as a result of treaty obligations, or for other historically unique reasons—that the rich nations have been willing (often reluctantly) to provide such aid at a relatively high level and on a continuing basis. Nonetheless, the existence of such cases does suggest that throwing money at a problem can in some important instances ameliorate it even if not solve it entirely.

Polynesia, then, provides an unparalleled opportunity to examine the health consequences of European contact over a period of two centuries. The historical picture has turned out to be complex: neither as uniformly catastrophic as older interpretations had it, nor as benign as the new Pacific history suggests. Similarly, the recent past and the present are also surprisingly complex. For all its undoubted injustices, economic growth has led to unrivalled improvements in life expectancy in Hawaii. But so has economic non-development in certain circumstances. Neither biological, social, nor economic theories alone can adequately explain the evolution of population and mortality over the past 200 years.

APPENDIX

Estimates of various Polynesian populations: 1790s–1980s

Decade	Tongans (1)	Samoans (1) American	Western	Total	Maoris (2)	Hawaiians (3)	Polynesians (4) French Tahitians	Marquesans
1790s					110 000	250 000	16–50 000	
1800s								
1810s								
1820s				40 000		145 000	8600	
1830s								
1840s	>18 500				80–90 000		8000	19 300
1850s	20–30 000			33 900	58–65 000			11 900
1860s	20–25 000					71 000	7000	7411
1870s					47–49 000	51 500		6000
1880s				35 000	43 900	44 200	9200	5200
1890s	19 196				42 100	39 500		4200
1900s	20 019			40 000	50 300	37 600	11 100	3500
1910s	21 712	7251	35 404	42 655	52 800	38 500	11 300	3100
1920s	23 759	8056			63–70 000	41 700	11 700	2300
1930s	27 700	10 055	52 266	62 321	83 300	50 800	16 700	2300
1940s	32 862	12 908			98 700	64 300	23 100	2700
1950s	55 156	20 154	91 833	117 401	137 100	86 000	30 500	3257
1960s					167 000	102 400	62 000	
1970s					270 035		80 000	
1980s	95 200	32 400	155 000	187 400	279 255	118 251	119 000	5419

(1) McArthur (1968); Taylor et al. (1989).
(2) Pool (1977, persl comm. 1991); Pomare and de Boer (1988). The first date for which estimates are available is 1769.
(3) Nordyke (1989).
(4) McArthur (1968); Vigneron (1989), for Marquesans in the 1970s, see Institute National de la Statistique et des Etudes Economique (1977). Vol. 1, p. 21.

ACKNOWLEDGEMENTS

The advice and comments of Gavan Daws, Donald Denoon, Alan Howard, Jerrold E. Levy, Cluny Macpherson, Ian Pool, and Gigi Santow are gratefully acknowledged. Douglas Crews, Kamuta Seuseu, and Charles McCuddin made available unpublished material, for which I thank them.

NOTES

1   There is some reason to believe that there may be more diversity in the biological inheritance of Polynesians than is commonly assumed, an issue beyond the bounds of this chapter (Langdon 1988).

2   The Appendix contains the numbers on which Fig. 3.1 is based, as well as the sources from which they are taken. All the sources are secondary; that is to say, I have used estimates made by contemporary historical demographers who have themselves analysed the primary data.

3   There has been, in fact, a reawakening of Hawaiian self-consciousness and pride during the past generation. Whether it compares in intensity, breadth, and depth to the Maori renaissance referred to by Beaglehole is beyond the scope of this chapter.

4   In 1900, the urban population included only Honolulu. In 1910 and 1920 it included Hilo as well.

5   Honolulu County includes the entire island of Oahu, but the city of Honolulu accounts for most of the population. The counties other than Honolulu began to grow rapidly in the 1970s with the growth of the tourist industry, which, since World War II had replaced sugar as a major source of income.

6   The definition of 'Hawaiian' and 'Maori' has been a troublesome issue for vital statisticians, demographers, and presumably for many Maoris and Hawaiians as well. Until the 1970 census Hawaiians have been classified as either Hawaiian or part-Hawaiian depending upon parentage (whether full or part-Hawaiian ancestry). Starting in 1970 the definition was changed to depend upon self-definition. The Hawaiian Health Survey of the State Health Department continues to use the initial census definition. The difference in the number of Hawaiians is substantial; far fewer people define themselves as Hawaiian in the census than are defined as Hawaiian in the Health Department Survey (Nordyke 1989, p. 104). The data I have used are based upon the Health Department definition.

The Maori life expectancy data are based upon the so-called 'biological' criterion of all those people who are at least half Maori (Pomare and de Boer 1988, pp. 25–6). There is some reason to believe that even this criterion must include a high proportion of people with less than half Maori ancestry (Pool

1977, p. 46). In 1981, there were 385 224 people who claimed some Maori ancestry compared with 279 255 who claimed to be half or more Maori in origin.

Thus, both the Hawaiian and Maori life expectancies are based upon a 'biological' measure having to do with parentage rather than upon a measure dependent upon 'self-definition'. On the other hand, the degree of Maori ancestry is more restrictive than the degree of Hawaiian ancestry. It is possible that these differences account for some of the differences in life expectancy reported for recent decades. They are less likely to have confounded the patterns in the inter-war and immediate post-war years when intermarriage was much less frequent (see, for example, Howard 1974, pp. 21–2).

7 Selective migration does not seem to adequately explain the difference.

8 Each of these groups seems to engage in more high-risk behaviours than non-indigenous peoples in the same countries (e.g. Chung et al. 1990; Sachdev 1990).

REFERENCES

Baker, P. T. and Crews, D. E. (1986). Mortality patterns and some biological predictors. In *The changing Samoans: behaviour and health in transition* (ed. P. T. Baker, J. M. Hanna, and T. S. Baker). Oxford University Press.

Beaglehole, E. (1939). *Some modern Hawaiians*. Research Publications No. 19. University of Hawaii Press, Honolulu.

Belshaw, H. (1940). Economic circumstances. In *The Maori people today: A general survey* (ed. I. L. G. Sutherland). Whitcombe and Tombs, Christchurch.

Bertram, I. G. and Watters, R. F. (1985). The MIRAB economy in South Pacific microstates. *Pacific Viewpoint*, **26**, 497–519.

Betram, I. G. and Watters, R. F. (1986). The MIRAB process: earlier analysis in context. *Pacific Viewpoint,* **27**, 47–59.

Bureau of the Census (1922). *Fourteenth census of the United States taken in the year 1920, Vol. III. Population 1920. Composition and characteristics of the population by states*. Government Printing Office, Washington, DC.

Bureau of the Census (1984). *Detailed social and economic characteristics. American Samoa.* PC80-1-C/D 56. US Government Printing Office, Washington, DC.

Bureau of the Census (1988). *Asian and Pacific islander population in the United States: 1980.* PC80-2-1E. US Government Printing Office, Washington, DC.

Carroll, V. (1975). The population of Nukuoro in historical perspective. In *Pacific atoll populations* (ed. V. Carroll). The University of Hawaii Press, Honolulu.

Cassel, J. (1974). Hypertension and cardiovascular disease in migrants: a potential source for clues. *International Journal of Epidemiology*, 3, 204–6.

Cassel, J., Patrick, R., and Jenkins, D. (1960). Epidemiological analysis of the health implications of culture change: a conceptual model. *Annals of the New York Academy of Science*, **84**, 938–49.

Chung, C. S., Tash, E., Raymond, J., Yasunobu, C., and Lew, R. (1990). Health risk behaviours and ethnicity in Hawaii. *International Journal of Epidemiology*, **19**, 1011–8.

Crews, D. E. (1985). Mortality, survivorship and longevity in American Samoa, 1950 to 1981. Unpublished Ph.D. dissertation. Department of Anthropology, The Pennsylvania State University.

Crews, D. E. (1988*a*). Body weight, blood pressure and the risk of total and cardiovascular mortality in an obese population. *Human Biology*, **60**, 417–33.

Crews, D. E. (1988*b*). Multiple cause of death and the epidemiological transition in American Samoa. *Social Biology*, **35**, 198–213.

Crews, D. E. (1989). Multivariate prediction of total and cardiovascular mortality in an obese Polynesian population. *American Journal of Public Health*, **79**, 982–6.

Crews, D. E. and Pearson, J. D. (1988). Cornell Medical Index responses and mortality in a Polynesian population. *Social Science and Medicine*, **27**, 1433–7.

Crocombe, R. (ed.) (1987). *Land tenure in the Pacific* (rev. edn). University of the South Pacific Press, Suva.

Crosby, A. W. (1986). *Ecological imperialism: The biological expansion of Europe 900–1900*. Cambridge University Press.

Daws, G. (1967). Honolulu in the 19th century: Notes on the emergence of urban society in Hawaii. *The Journal of Pacific History*, **2**, 77–96.

Denoon, D. (1983). *Settler capitalism*. Oxford University Press.

Department of Health (1990). *Annual Report of the Department of Health, 1986*. Western Samoa Government Printing Office, Apia.

Department of Statistics (1982). *New Zealand census of population and dwellings, 1981. Vol. 8, Part A. New Zealand Maori population and dwellings*. Department of Statistics, Wellington.

Department of Statistics (1988). *Annual statistical abstract. Western Samoa*, Vol. 23. Western Samoa Government Printing Office, Apia.

Eggan, F. (1954). Social anthropology and the method of controlled comparison. *American Anthropologist*, **56**, 743–63.

Fitzgerald, M. H. and Howard, A. (1990). Aspects of social organization in three Samoan communities. *Pacific Studies*, **14**, 31–54.

Fleischman, R. K. (1982). *Death in Paradise: Big island mortality 1910–1950*. R&S Report, No. 40. Hawaii State Department of Health, Honolulu.

Freeman, D. (1983). *Margaret Mead and Samoa: The making and unmaking of an anthropological myth*. Harvard University Press, Cambridge, MA.

Gardner, R. W. (1980). Ethnic differentials in mortality in Hawaii, 1920–1970. *Hawaii Medical Journal*, **39**, 221–6.

Gardner, R. W. (1984). *Life tables by ethnic group for Hawaii, 1980*. R&S Report, No. 47. Hawaii State Department of Health, Honolulu.

Garwood, A. N. (ed.) (1986). *Almanac of the 50 states*. Information Publishers, Newburyport, MA.

Hamlin, H. (1932). The problem of depopulation in Melanesia. *Yale Journal of Biology and Medicine*, **41**, 301–21.

Hanna, J. M., Pelletier, D. L., and Brown, V. J. (1986). The diet and nutrition of contemporary Samoans. In *The changing Samoans: Behaviour and health in transition* (ed. P. T. Baker, J. M. Hanna, and T. S. Baker). Oxford University Press.

Hayes, G. (1991). Migration, metascience, and development policy in island Polynesia. *The Contemporary Pacific*, **3**, 1–58.

Hecht, J. A., Orans, M., and Janes, C. R. (1986). Social settings of contemporary Samoans. In *The changing Samoans: Behaviour and health in transition* (ed. P. T. Baker, J. M. Hanna, and T. S. Baker). Oxford University Press.

Heffernan, L. (1988). From independent nation to client state: the metamorphosis of the Kingdom of Hawai'i in the pages of the *North American Review* in the nineteenth century. *The Hawaiian Journal of History*, **22**, 209–27.

Heighton, R. H., Jun. (1968). Physical and dental health. In *Studies in a Hawaiian community: Na Makamaka o Nanakuli* (ed. R. Gallimore and A. Howard). Pacific Anthropological Records, No. 1, Bernice P. Bishop Museum, Honolulu.

Holmes, L. D. (1987). *Quest for the real Samoa: The Mead/Freeman controversy and beyond*. Bergin and Garvey, South Hadley, MA.

Howard, A. (1974). *Ain't no big thing: Coping strategies in a Hawaiian–American community*. University of Hawaii Press, Honolulu.

Howard, A. (1986). Samoan coping behaviour. In *The changing Samoans: Behaviour and health in transition* (ed. P. T. Baker, J. M. Hanna, and T. S. Baker). Oxford University Press.

Howe, K. R. (1984). *Where the waves fall*. University of Hawaii Press, Honolulu.

Institut National de la Statistique et des Etudes Economiques (1977). *Résultats du recensement de la population de la Polynésie Française, 29 Avril 1977*. Paris, France.

Janes, C. R. (1990). *Migration, social change, and health: A Samoan community in urban California*. Stanford University Press.

Keesing, F. M. (1945). *The South Seas in the modern world*. The John Day Company, New York.

Kent, N. J. (1983). *Hawaii: Islands under the influence*. Monthly Review Press, New York.

Kunitz, S. J. (1983). *Disease change and the role of medicine*. University of California Press, Berkeley.

Kunitz, S. J. (1984). Mortality change in America, 1620–1920. *Human Biology*, **56**, 559–82.

Kunitz, S. J. (1990). *Disease and the destruction of indigenous populations*. Working paper No. 15. National Centre for Epidemiology and Population Health, Australian National University, Canberra.

Lambert, S. M. (1934). *The depopulation of Pacific races*. Bernice P. Bishop Museum Special Publication, No. 23, Honolulu.

Lambert, S. M. (1941). *A Yankee doctor in paradise*. Little, Brown, Boston.

Langdon, R. (1988). *The lost caravel re-explored*. Brolga Press, Canberra.

Le Marchand, L. and Kolonel, L. N. (1989). Cancer: epidemiology and prevention. *Social Process in Hawaii*, **32**, 134–48.

Lind, A. W. (1930). Some ecological patterns of community disorganization in Honolulu. *American Journal of Sociology*, **36**, 206–20.

Lind, A. W. (1931). The ghetto and the slum. *Social Forces*, **9**, 206–15.

Lind, A. W. (1938). *An island community*. University of Chicago Press.

Look, M. A. (1982). *A mortality study of the Hawaiian people*. R&S Report, No. 38. Hawaii State Department of Health, Honolulu.

Macpherson, C. amd Macpherson, L. (1987). Towards and explanation of recent trends in suicide in Western Samoa. *Man* (N.S.), **22**, 305–30.

Macpherson, C. and Macpherson, L. (1990). *Samoan medical belief and practice*. Auckland University Press.

McArthur, N. (1968). *Island populations of the Pacific*. Australian University Press, Canberra.

McArthur, N., Saunders, I. W. and Tweedie, R. L. (1976). Small population isolates: a micro-simulation study. *Journal of the Polynesian Society*, **85**, 307–26.

McCreary, J. R. (1968). Population growth and urbanisation. In *The Maori people in the Nineteen-Sixties* (ed. E. Schwimmer). Blackwood and Janet Paul Ltd, Auckland.

McCuddin, C. (1989). *Overview of American Samoa health care system and health status of the population*. Division of Planning and Development, Department of Health, L.B.J. Tropical Medical Center, Pago Pago.

McGarvey, S. T. and Schendel, D. E. (1986). Blood pressure of Samoans. In *The changing Samoans: Behaviour and health in transition* (ed. P. T. Baker, J. M. Hanna, and T. S. Baker). Oxford University Press.

McNeil, W. M. (1976). *Plagues and peoples*. Penguin, Harmondsworth.

Mead, M. (1928). *Coming of age in Samoa*. Morrow, New York.

Mead, M. (1930). *Social organization of Manu'a*. Bernice P. Bishop Museum, Honolulu.

Meleisea, M. (1987). *The making of modern Samoa: Traditional authority and colonial administration in the modern history of Western Samoa*. University of the South Pacific Press, Suva.

Moorehead, A. (1968). *The fatal impact: an account of the invasion of the South Pacific 1767–1840*. Penguin, Harmondsworth.

Native Hawaiians Study Commission (1983). *Report on the culture, needs and concerns of Native Hawaiians*, Vols I and II. US Department of the Interior, Washington, DC.

Newbury, C. (1980). *Tahiti Nui: Change and survival in French Polynesia 1767–1945*. University of Hawaii Press, Honolulu.

Ngata, A. (1940). Maori land settlement. In *The Maori people today: A general survey* (ed. I. L. G. Sutherland). Whitcombe and Tombs, Christchurch.

Nordyke, E. C. (1989). *The peopling of Hawaii*. The University of Hawaii Press, Honolulu.

O'Meara, J. T. (1990). *Samoan planters: Tradition and economic development in Polynesia*. Holt, Rinehart and Winston, Fort Worth, Texas.

Oliver, D. (1981). *Two Tahitian villages: A study in comparison*. The Institute for Polynesian Studies, Brigham Young University, Laie, Hawaii.

Omran, A. R. (1971). The epidemiologic transition: A theory of the epidemiology of population change. *Milbank Memorial Fund Quarterly*, **49**, 509–38.

Pearson, D. (1990). *A dream deferred: The origins of ethnic conflict in New Zealand*. Allen and Unwin, Wellington.

Pirie, P. (1972). Population growth in the Pacific islands: the example of Western Samoa. In *Man in the Pacific Islands* (ed. R. G. Ward). Oxford University Press.

Pitt-Rivers, G. H. L-F. (1927). *The clash of culture and the contact of races.* George Routledge and Sons, London.

Pomare, E. (1980). *Maori standards of health: A study of the 20 year period 1955–1975.* Special Report Series, No. 7. Medical Research Council of New Zealand, Wellington.

Pomare, E. and de Boer, G. (1988). *Hauora: Maori standards of health: A study of the years 1970–1984.* Special Report Series, No. 78. Medical Research Council of New Zealand, Wellington.

Pool, D. I. (1977). *The Maori population of New Zealand, 1769–1971.* University of Auckland Press.

Pool, D. I. (1985). *Population of New Zealand,* Vol. 1. Country Monograph Series, No. 12. United Nations Economic and Social Commission for Asia and the Pacific, Bangkok, Thailand.

Prior, I. A. M. (1968). Health . In *The Maori people in the Nineteen-Sixties* (ed. E. Schwimmer). Blackwood and Janet Paul Ltd, Auckland.

Rallu, J-L. (1990). *Les populations océaniennes au Xix^e et Xx^e siècles.* Institut National d'Etudes Demographiques, Paris.

Ralston, C. (1977). *Grass huts and warehouses: Pacific beach community of the nineteenth century.* Australian National University Press, Canberra.

Redfield, R. (1955). *The little community: Viewpoints for the study of the human whole.* The University of Chicago Press.

Rivers, W. H. R. (ed.) (1922). *Essays on the depopulation of Melanesia.* Cambridge University Press.

Roberts, S. H. (1927). *Population problems of the Pacific.* George Routledge and Sons, London.

Sachdev, P. S. (1990). Behavioural factors affecting physical health of the New Zealand Maori. *Social Science and Medicine,* **30,** 431–40.

Scarr, D. (1990). *The history of the Pacific Islands: Kingdoms of the reef.* Macmillan, South Melbourne.

Schmitt, R. C. (1968). *Demographic statistics of Hawaii, 1778–1965.* University of Hawaii Press, Honolulu.

Schoeffel, P. (1984). Dilemmas of modernization in primary health care in Western Samoa. *Social Science and Medicine,* **19,** 209–16.

Stannard, D. (1989). *Before the Horror: The population of Hawai'i on the eve of Western contact.* Social Science Research Institute, University of Hawaii, Honolulu.

Sunia, I. F. (1983). American Samoa: Fa'a Amerika? In *Politics in Polynesia.* University of the South Pacific Press, Suva.

Taylor, R., Lewis, N. D., and Levy S. (1989). Societies in transition: mortality patterns in Pacific island populations. *International Journal of Epidemiology,* **18,** 634–46.

Vayda, A. (1959). Polynesian cultural distributions in new perspective. *American Anthropologist,* **61,** 817–28.

Vigneron, E. (1989). The epidemiological transition in an overseas territory: disease mapping in French Polynesia. *Social Science and Medicine,* **29,** 913–22.

Wegner, E. L. (1989). Hypertension and heart disease. *Social Process in Hawaii*, **32**, 113–33.

Zimmet, P., Faaiuso, S., Ainuu, J., Whitehouse, S., Milne, B., and De Boer, W. (1981). The prevalence of diabetes in the rural and urban Polynesian population of Western Samoa. *Diabetes*, **30**, 45–51.

INDEX